The Unity Game Engine and the Circuits
of Cultural Software

Benjamin Nicoll · Brendan Keogh

The Unity Game Engine and the Circuits of Cultural Software

palgrave
macmillan

Benjamin Nicoll
Digital Media Research Centre,
School of Communication
Queensland University of Technology
Brisbane, QLD, Australia

Brendan Keogh
Digital Media Research Centre,
School of Communication
Queensland University of Technology
Brisbane, QLD, Australia

ISBN 978-3-030-25011-9   ISBN 978-3-030-25012-6   (eBook)
https://doi.org/10.1007/978-3-030-25012-6

Cover illustration: © John Rawsterne/patternhead.com
Cover design by eStudioCalamar

This Palgrave Pivot imprint is published by the registered company Springer Nature
Switzerland AG
The registered company address is: Gewerbestrasse 11, 6330 Cham, Switzerland

ACKNOWLEDGEMENTS

This book represents a synthesis of two separate research projects, managed independently by both authors. Benjamin Nicoll thanks the University of Melbourne's Networked Society Institute, Intellectual Property Research Institute of Australia, and Centre for Media and Communications Law for jointly funding two collaborative research grants on game engines, which supported his research for this book in 2018. He would also like to thank Megan Richardson, Bjorn Nansen, Adam Lodders, and Jeannie Paterson for helping to secure this financial support, and for providing indispensable mentorship and opportunities for collaboration; Dan Golding, Dale Leorke, and Helen Stuckey for their collegial contributions and suggestions with regard to the above grants; John Sietsma and Chris Murphy for offering crucial insights into the world of commercial game engines; and Ceri Hutton, who was generous in accommodating his research residency at The Arcade. Benjamin also thanks his interview participants for taking the time to be interviewed, for agreeing to be quoted in publications based on the research, and for their willingness to share their thoughts and opinions on game engines. Beyond these acknowledgements, Benjamin thanks his partner, Britt, for a contribution beyond measure.

Brendan Keogh's contribution to this book was supported by the Australian Research Council Discovery Early Career Fellowship, 'Australian Game Developers and Skills Transfer' (DE180100973), and he wishes to thank all his research participants for their insights, time, and candour. He is also grateful to his friends Dan Golding,

Ben Abraham, and Terry Burdak for their ongoing support. Finally, Brendan thanks his partner, Helen Berents, not only for her endless support but also in this case for her expert feedback and suggestions on how we depict the circuits of our cultural software framework and, as an International Relations scholar, on our use of the concepts of 'governance' and 'democracy'.

Both authors would like to thank Mala Sanghera-Warren and Lucy Batrouney, our editors at Palgrave Macmillan, for their guidance and support throughout the proposal, peer review, and publication process. We would also like to thank the anonymous reviewers for their incredibly thoughtful, detailed, and encouraging comments on the manuscript.

The authors received permission to publish Fig. 1.1 from Grace Bruxner; Fig. 3.1 from Unity Technologies; and Fig. 6.1 from Bennett Foddy. Unity and Unity logos are registered trademarks and trademarks of Unity Technologies or its affiliates in the USA and elsewhere. All rights reserved. Other images, content, names, or brands are proprietary of their respective owners. Neither this book nor its authors are affiliated with, or endorsed or sponsored by, Unity Technologies or its affiliates.

CONTENTS

LIST OF FIGURES

CHAPTER 1

The Unity Game Engine and the Circuits of Cultural Software

Abstract This chapter describes the 'circuits of cultural software' as a framework that guides the book and its analysis; offers a preliminary definition of game engines; and introduces the Unity game engine as the book's core case study. It also discusses key terms such as cultural software, proprietary and commercial game engines, workflow, grain, literacy, and governance, and situates the book in relation to existing research on videogame production, game engines, and software culture. It briefly discusses Unity's place in Australia's videogame industry—which is where the research for the book was conducted—and provides a chapter outline.

Keywords Cultural software · Unity game engine · Circuit of culture · Game engine · Software studies · Platform studies

The videogame *Grace Bruxner Presents: The Haunted Island, a Frog Detective Game* (Bruxner, 2018) is notable in its simplicity. It is approximately one hour long, and is premised on exploration, observation, and reading rather than complex systems, challenges, and goals (see Fig. 1.1). Its charming visual style and clever writing have seen it nominated for a number of awards at international videogame festivals, and it has received extensive coverage in the videogame press. Yet, *The Haunted Island* was not made in a typical videogame development environment—that is, in a studio comprised of large groups of specialist creative workers and

B. Nicoll and B. Keogh, *The Unity Game Engine and the Circuits of Cultural Software*, https://doi.org/10.1007/978-3-030-25012-6_1

Fig. 1.1   The detective inspects a bowl of pasta in *Grace Bruxner Presents: The Haunted Island, a Frog Detective Game* (Bruxner, 2018). By permission of Grace Bruxner

corporate resources. It was developed primarily by one person, Grace Bruxner, with programming and audio support from Tom Bowker and Dan Golding, respectively. Grace wrote the dialogue, modelled and animated the characters, designed the layout of the virtual world, and put together the videogame's events. Notably, Grace was able to make *The Haunted Island* while still completing a videogame design undergraduate degree at RMIT University in Melbourne. To do this, Grace took advantage of a commercial software tool known as Unity, owned by Unity Technologies.[1] Without paying any fees upfront, and without the need for low-level computer science skills, Grace used Unity to put together *The Haunted Island*'s necessary elements and export 'builds' for Windows and Mac.

Today's videogame-making ecology is increasingly inhabited by creators who, like Grace, are taking advantage of low-cost and low barrier to

---

[1]Where feasible, throughout this book we use 'Unity Technologies' to refer to the company and 'Unity' to refer to the game engine owned by that company. However, in popular vernacular, and thus in many of our respondents' quotes, Unity Technologies is often referred to as 'Unity'.

entry software tools to produce a wide range of videogame works. Many of these creators are no longer confined to traditional studio environments, and are instead working across a spectrum of formal and informal contexts. Several cultural and technical factors have afforded this diffusion (see Keogh 2019), but in this book we are centrally concerned with Unity. Unity is a software development tool commonly identified as a 'game engine'. Games engines enable programmers, designers, and artists to build, collaborate on, and run real-time interactive digital content, including (but not limited to) videogames. In videogame development, game engines function as software hubs wherein a vast range of media forms and skills converge into singular videogame builds. Game engines have been foundational to videogame development since at least the mid-1990s, yet the last decade has seen radical shifts in the availability and accessibility of various game engines, each with their own affordances. This, in turn, has created fertile ground for a plurality of videogame styles, genres, and developer identities to emerge, in a manner not dissimilar to the introduction of the Kodak camera or the 8-track tape. Unity holds a notable position in these shifts. Its low-cost availability, relative ease of use, and ability to scale to a vast range of student, amateur, professional, and industrial applications have seen it come to dominate videogame production globally, to such an extent that the CEO of Unity Technologies, John Riccitiello, boasts that over half of all videogame and virtual reality projects on contemporary devices are developed in Unity (Dillet 2018).

Game engines are typically owned and distributed by commercial companies that are directly invested in ensuring their engines capture a large market share. Unity, with its accessible editing interface, flexible licensing structure, and modular toolset, is framed by company representatives as an almost revolutionary piece of software that is 'democratizing game development' and 'empowering game developers' (see Unity 2018). To this end, Unity is associated with a levelling out of work role hierarchies in studio environments—hierarchies that, historically, have delegated power to programmers and software engineers as opposed to artists and designers such as Grace. Yet, while Unity claims to have democratized the means of videogame production, it has also provoked the ire (and, in some cases, outright hatred) of a small—yet vocal—group of developers, critics, and players. A brief search on any videogame enthusiast discussion board yields accusations that Unity's accessibility is causing an oversaturation of low-quality videogames, a dearth of programming skills, and a proliferation of 'asset flipping' in videogame development—a derogatory expression referring to videogames constructed from prefabricated (i.e. store-bought) parts or assets (Grayson 2018). In a similar vein, some developers perceive a looming 'indiepocalypse' of supply overwhelming demand as a repeat of the North American videogame industry crash of 1983, which almost destroyed Atari and a national industry (Pedercini 2017). Some industry professionals and educators express concern that junior developers and students are not 'really' learning how to make videogames, but simply learning how to use Unity. Digital marketplaces, such as Valve's Steam platform, have made public promises to crack down on ostensibly 'fake games' made in Unity. Scholars, too, express concern that game engines have, since their introduction in the 1990s, led to a homogenization and rationalization of videogame production (Kirkpatrick 2013: 105–106; see also Freedman 2018a: n.p.). Game engines can also be understood in terms of a broader 'platformization of cultural production' (Nieborg and Poell 2018), wherein cultural production is increasingly controlled by a small number of dominant platform companies. These varied anxieties point to a radical reconfiguration of the practices, identities, values, and contexts associated with videogame development today.

How, then, might we make sense of these cultural, technological, and design shifts that, at once, seem to empower developers such as Grace, yet that also seem to make developers beholden to a single company's product? It is this duality that this book is centrally concerned with. In the chapters that follow, we argue that game engines are a form of *cultural software*, and that their social, political, technological, and ideological effects must be mapped and analysed. While Lev Manovich (2013: 21) defines cultural software as 'software that support actions we normally associate with culture', we adopt a narrower definition: *cultural software are software that provide code frameworks for actions we normally associate with cultural production*. We are thinking, here, of software tools such as Photoshop, Blender, Garage Band, Final Cut Pro, and, of course, Unity. Such programs, we argue, enrol their users in *circuits of cultural software* in the way they influence, mediate, and articulate the processes and contexts of cultural production. Cultural software have a fundamental bearing on production *workflows* across different design contexts. They encourage media creatives to adopt particular design methodologies and thus possess varying *grains—* protocols, standards, and affordances—that give shape to creative expression. Cultural software promote the cultivation of specific *literacies* in their respective areas of cultural production—not simply through inbuilt tutorials, but also through their embeddedness in company-specific development environments, educational contexts, and the 'collective intelligence' (Lévy 1997; cf. Jenkins 2006) of online communities. Cultural software deploy platform-based business models and policy discourses to strategically *govern* the activities of their users. Through an analysis of Unity specifically and game engines more generally, this book makes quite a simple argument: game engines are more than just 'actors' situated in studio environments. They are also cultural software whose articulations within and across a number of interconnected cultural circuits now need to be taken into account.

THE CIRCUITS OF CULTURAL SOFTWARE

In the following chapters, we develop a framework for understanding and articulating the effects of cultural software on the process of cultural work, which we call the circuits of cultural software.[2] This framework has obvious affinities with the influential 'circuit of culture' approach (du Gay et al. 1997), in which a cultural object (the prototypical example being the Sony Walkman) is passed through five interlinked sites—representation, identity, production, consumption, and regulation—and researched accordingly. The circuit of culture illustrates that cultural objects gain meaning not only through processes of production and consumption but also through their representation and articulation in symbolic and discursive contexts. However, the circuit of culture cannot be applied wholesale to today's software-based culture. Software is the 'engine' of twenty-first-century cultural production (Manovich 2013), just as industrialized mass production was the 'engine' of cultural production in the mid-twentieth century (Adorno and Horkheimer 2002 [1947]). For Manovich (2013: 33), software tools such as Photoshop,

---

[2]We are not the first to utilize traditional cultural studies approaches and methods to discuss and analyse how cultural software produce and circulate meaning. See, for example, Zhao et al. (2014) for an analysis of PowerPoint as a 'semiotic technology'. See also Kline et al. (2003), for a 'three circuits of interactivity' model that synthesises methods from cultural studies, media studies, and media theory to analyse the concept of 'interactivity' and its articulations (and contradictions) within and across videogame technology, culture, and marketing.

Blender, and Maya 'play a central role in shaping both the material elements and many of the immaterial structures that together make up "culture"', and so configure the very circuitry that underpins capital, labour, and creativity in today's economy. In software culture, surplus value[3] is generated from an interplay between informal and formal modes of human capital (Qiu et al. 2014; Lobato and Thomas 2015; Keogh 2019), in a way that is consonant with a broader neoliberalization of work and subjectivity (Chun 2011). The traditional circuit of culture, with its 'free flowing, even idealistic' structure, has difficulty accounting for these in/formal flows of human capital and creativity (Qiu et al. 2014: 568). More fundamentally, as Lawrence Grossberg (1997: 256) argues, 'one cultural studies investigation is not the same as that of another'—a fundamental truism not reflected in the non-specificity of the traditional circuit of culture.

Our framework (Fig. 1.2), therefore, is not meant as a replacement for the original circuit of culture, but rather as a particular instantiation of the circuit of culture for our current cultural period. A key difference is that we are not tracing a single cultural object but rather oftentimes opaque software frameworks (such as Unity) upon which cultural objects (such as videogames) are typically produced, and out of which various cultural scenes, aesthetics, and discourses emerge. As such, the cultural software being analysed—whether that be Unity, Photoshop, Maya, or whatever—do not figure in the framework, but are instead *constitutive of the framework*. In its default state (having not yet been articulated to a particular cultural software), our framework is comprised of three overlapping circuits of mediation—*workflow*, *grain*, and *literacy*—encircled and permeated by a broader *governance* circuit. *Workflow* refers to the ways in which cultural software position themselves as 'meta-platforms' (Bratton 2015: 65) for the coordination of intensely individualized labour and production processes. *Grain* refers to the design methodologies that cultural software orient their users towards. *Literacy* refers to the ways of knowing and identifying that come to be associated with specific cultural software. The broader *governance* circuit refers

---

[3] 'Surplus value' is a term that originates in the writings of Karl Marx, and is generally used to refer to capital generated from a combination of (a) human labour, which is exploited to produce commodities, and (b) the market value of said commodities, which far exceeds the costs associated with exploited labour. The resulting surplus value is redirected to wealthy capitalists, thus reinforcing a structure of power that subordinates workers.

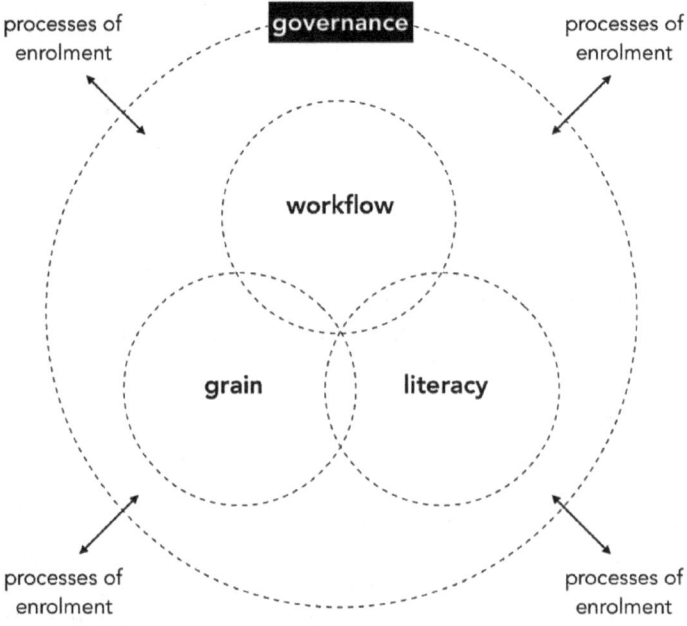

Fig. 1.2   Circuits of cultural software

to the encompassing policy[4] discourses and modes of affective inter-mediation that cultural software deploy in order to *enrol* potential constituents into their software ecologies. These processes of enrolment are represented by the arrows pointing into and out of the governance circuit. We understand enrolment similarly to James Ash's (2015) concept of the 'interface envelope' or Thomas Apperley et al.'s (2016) concept of the 'aesthetics of recruitment'. That is, cultural software recruit users not only through their immediate interfaces, but also by monopolizing their interface effects to encompass the 'radically contextual' (Grossberg 1995, 2010) 'postdigital' (Berry and Dieter 2015) environments in which they are situated. Potential users, audiences, publics, and other stakeholders are enrolled into the circuits as *constituents.*

---

[4]We use the term 'policy' to refer not only to Unity's policies as a company—its licensing structures and its terms and conditions, for example—but also its policy platform, which is articulated through marketing slogans such as 'democratizing game development'.

In turn, constituents enrol cultural software into their existing workflows and design methodologies—refashioning their practitioner identities and software literacies accordingly. This process of enrolment flows in both directions; constituents can be enrolled, but they can also 'short-circuit' the software ecology for their own ends—that is, they can 'absorb productive energy' (Qui et al. 2014: 564) from any given software ecology for potentially transgressive or countercultural purposes. Our conception of the circuits of cultural software is similar to Benjamin H. Bratton's (2015: xviii) concept of the 'software stack'—an 'accidental megastructure' wherein 'hard and soft' tools, techniques, protocols, social forces, and human actors 'operate within a modular and interdependent vertical order', amounting to a computational form of governmentality whose remit extends far beyond traditional notions of state sovereignty and, indeed, culture.

In the pages that follow, we observe Unity's articulations within and across the circuits of cultural software. In terms of workflow, we find that Unity promotes a component-oriented design system; that it decentres the role of the programmer, who typically sits atop the software development hierarchy; and that it operates as a site of 'deep remixability' (Manovich 2013: 268) for differentiated software techniques. When it comes to grain, we find that videogames produced in Unity often possess an ineffable 'look and feel' through the processes and frameworks that Unity presents as either defaults or preferences. When it comes to literacy, we find that different constituents understand Unity's mediations as empowering, diversifying, homogenizing, or threatening to the videogame field. Unity mobilizes each of these circuits—workflow, grain, and literacy—to enrol constituents into its circuit of governance, where, under the guise of what we call a 'democratization *dispositif*',[5] people are encouraged to *self*-govern and become entrepreneurs of themselves (cf. Foucault 2008: 226). It is important to note, here, that videogame-making workflows, methodologies, literacies, and governance structures existed before Unity and will continue to exist after Unity. Moreover, in forwarding such a framework, we do not mean to position specific software programs as central actors around which a cultural field

---

[5]In using this expression, we are building on Angela McRobbie's (2016: 86) notion of a 'creativity *dispositif*', which refers to the 'toolkits, instruments and new entrepreneurial pedagogies' that encourage prospective media creatives to embark on careers in the risk economy of creative work.

orbits. Rather, by figuring cultural software *as* these interconnected circuits, we are interested in how cultural software mobilize, support, and make serviceable these circuits for their own ends, by bringing constituents into contact with their governance structures.

Our analysis is driven by three core questions: first, *what does Unity do within and across the circuits of cultural software*, in both a technical and cultural sense? Second, *how do videogame developers discover Unity and become enrolled in its software ecology?* A videogame developer may discover Unity by, for example, starting with the core software, establishing a workflow and coming to grips with the engine's grain, before identifying their own niche within the engine's software ecology. Finally, we also ask *how Unity discovers and enrols potential Unity constituents?* Someone may discover Unity by first encountering its slogan, 'democratizing game development', or by playing various Unity-developed videogames, or by hearing it decried in an online editorial lamenting the glut of 'fake games'. Indeed, one of the key reasons for Unity's success—and something that will be discussed in more detail in relation to the engine's governance circuit—is its ability to make videogame development seem like an exciting, accessible, and viable career path. For this reason, the lines separating each of the circuits are not depicted as impenetrable or hierarchical but rather porous and flat. Taking influence from the original circuit of culture, we note that one could start anywhere in the circuits and ultimately end up enrolled in the governance circuit—such is the 'accidental' (Bratton 2015) logic of software culture.

WHAT IS A GAME ENGINE?

Game engines are complex entities that elude singular categorization (see Banks 2013: 44–45). A key argument of this book is that game engines are, to borrow Grossberg's (1995, 2010) term, 'radically contextual', in that they gain new meanings as they move within and across the circuits of cultural software described above. Given that there are many different types of game engines—engines that, moreover, perform distinct roles and functions depending on the contexts in which they are articulated—we will start here with a basic (albeit reductive) technical definition. A game engine is a software tool that enables real-time interactive digital content to be created, and a code framework that enables that content to run on different platforms that might include consoles, smartphones, and virtual reality devices. Note, here, that game engines are used to create 'real-time interactive digital content' rather than simply 'videogames'. Game engines emerged from and are primarily associated with the videogame industry, but they can be (and increasingly are) used to make non-videogame content and applications such as 3D animations, architectural models, complex training simulations, and interactive novels.[6] As discussed above, game engines are also cultural software that encourage users to utilize their individual labour as itself an 'engine' of capital and creativity. Crucially, like all cultural software, game engines offer the promise of self-sovereignty—they make us feel empowered to take control of our creative potentials through accessible drag-and-drop tools, dashboard interfaces, and customizable management techniques, all by concealing the true basis of their operability under the guise of 'user-friendliness' (Chun 2011).

Different game engines are optimized for different types of content creation and production workflows, but on a very basic (and again, predominantly technical) level, game engines are designed to manage 'low-level' computational tasks such as rendering, physics, and artificial intelligence, thereby freeing up developers to focus on 'higher-level' aspects of the design process. Many (though not all) of today's game engines are interoperable with a range of different cultural software, programming languages, production workflows, and middleware[7] and plugin capabilities. Unity is a paradigmatic example of this interoperability, or what Manovich (2013: 268) calls 'deep remixability'. An artist, for example, might create assets (2D and 3D models, visual effects, or textures) using a software tool such as Blender or Maya, before importing their creations into Unity. Likewise, a programmer might write code in a text editor such as Visual Studio and host it on version control software such as Git, before finally testing their scripts in Unity. In this way, game engines can be understood as software metaplatforms that, to borrow Bratton's (2015: 65) words, work

---

[6] See Freedman (2018b: 327–328) for a discussion of the use of Unity and Unreal in non-videogame (or videogame-adjacent) contexts, including NASA's various spacecraft simulators and educational software, and BMW's vehicle prototyping software.

[7] Middleware are specialised software tools that expand a game engine's underlying toolset. Examples include standalone art tools such as Autodesk Maya and Blender, music tools such as FMOD, and physics tools such as Havok. In other contexts, these tools may be considered standalone pieces of software rather than middleware, but once brought into contact with game engines, their role is commonly reconceptualized as one of expanding the engine's underlying toolset.

'to gather, support, and superimpose' the broader 'pipeline' of video-game development. Pipeline is the term given to the production chain of design concepts, assets, animations, scripts, audio, and more, each of which are 'fed into' the game engine and made interoperable through the engine's underlying code framework. Today's game engines also tend to encourage rapid iteration, meaning that they make it possible for developers to prototype content (and iterate on content that already exists) by switching seamlessly between building, testing, and editing modes. Game engines tend not to be discrete, unchanging, or totally black-boxed objects, but rather modular toolsets that can be customized to facilitate a variety of different design methodologies and project types.

Anyone with considerable expertise, time, and resources can build a custom game engine, though most developers today opt to use an existing commercial or proprietary game engine. In this book, we use 'commercial game engine' to describe third-party engines such as Unity, because they are available through public commercial contracts. We use 'proprietary game engine' to describe engines that have been custom-made (usually within studios) and patented by their creators. However, as will become clear in Chapter 2, the commercial-proprietary distinction is slippery. Most game engines are inseparable from—indeed, even synonymous with—intellectual property, end-user license agreements (EULAs), and the patenting of company-specific development pipelines (see Bogost 2006: 56–57; Nieborg and van der Graaf 2008). Likewise, many proprietary game engines are partially available for commercial use, in the form of player-oriented level editors and modding toolkits. There are also engines that were once considered proprietary but are now commercially available, such as the Unreal engine. 'Commercial' and 'proprietary' are nonetheless industry-standard terms for differentiating third-party engines (such as Unity) from engines that have been custom-developed and patented by their creators. It is worth acknowledging, however, that these terms do not adequately capture the breadth of videogame-making tools that exist (see Freedman 2018a). Twine, for example, is a free open-source HTML-based interactive fiction editor that has been widely appropriated by marginal videogame makers (Harvey 2014), while *Super Mario Maker 2* (Nintendo, 2019) for the Nintendo Switch is a commercial videogame with a player-oriented level editor embedded within it. Elsewhere, smaller tools made by individuals rather than large corporations, such as Adam Le Doux's Bitsy and Lexaloffle's Pico-8, allow for the creation and distribution of videogames within very specific technical restrictions. For the purposes of this book, we consider this broader range of player-oriented, fringe, and grassroots videogame-making tools as 'game engines', as they too are cultural software that provide code frameworks for videogame production.

In the existing research in this area, game engines are often described as social entities that facilitate interdisciplinary collaboration and knowledge exchange within and across team compositions. John Banks (2013: 53), for example, describes the game engine as a 'multiple object' whose role in the development process is to coordinate the—oftentimes incompatible and conflicting—knowledge boundaries of programmers, designers, artists, and producers. Likewise, Jennifer R. Whitson (2018a) argues that game engines and related middleware exert 'voodoo agency' over their users. Game engines are often ascribed anthropomorphic qualities by their users—they exhibit their own biases and preferences, much like any other member of the development team—and because of this, argues Whitson (2018a: 2327), they can act as scapegoats for roadblocks encountered in the development process. In this way, they function to 'corral the negative social effects that result when one member's contribution to the game refuses to work or to fit nicely within the larger structure, thus preserving [team] cohesion' (Whitson 2018a: 2327). For Banks and Whitson, game engines are complex technical and social 'actors' that perform distinct 'roles' within and across the various contexts they are articulated. Much of the existing research on game engines is based on data produced at a time when proprietary game engines, which were typically custom-made within studios and safeguarded therein, were the dominant means of videogame production (see, for example, O'Donnell [2014] and Banks [2013]). However, today's commercial game engines facilitate ecologies of production, consumption, and meaning-making that far exceed the confines of the traditional studio environment. Unity is a key barometer—indeed, catalyst—of this shift. It is widely celebrated (and, in some contexts, disparaged) for making videogame development much 'easier' than it once was (a notion problematized in Chapter 5); for providing everyone from students to studio developers with low-cost, low barrier to entry, and standardized videogame-making tools; and for fostering grassroots development scenes that scale online, offline, and inter-regional videogame-making cultures and communities.

WHAT IS UNITY?

Unity's origin story is something of a pastiche of the 'rags-to-riches' (and highly gendered) success stories often associated with tech start-ups. 'In the early 2000s', writes Jon Brodkin (2013: n.p.) in an online article on the company's history, 'three young programmers without much money gathered in a basement and started coding what would become one of the most widely used pieces of software in the videogame industry'. Those three young programmers were David Helgason, Joachim Ante, and Nicholas Francis. Ante and Francis met on a Mac OpenGl board in May of 2002, after Ante responded to Francis's request for advice on a shader tool he was building for a custom engine (Haas 2014: 4). Ante and Francis subsequently began collaborating on said shader tool and, thereafter, a custom game engine. Helgason then joined the project and was appointed CEO. The team then formed the company Over the Edge Entertainment (OTEE)—now Unity Technologies—and began hiring software engineers. Inspired by Apple's Final Cut Pro, OTEE's initial plan was to create a Mac-only development tool that could be used to create and publish interactive web-based 3D content.[8] Their main competition was Adobe's Flash—a development tool that, while popular, supported a limited range of scripting languages and graphics capabilities. Once Unity was in beta, OTEE used the engine to develop and publish a videogame called *GooBall* (Over the Edge Entertainment, 2005). Like the flagship titles of many game engines, *GooBall* was developed to advertise Unity's capabilities (Haas 2014: 7–8).

Unity 1.0 was released in June of 2005. It subsequently went through multiple software iterations and name changes before it became what it is today (version 2019.1 at the time of writing). Its adoption rate grew dramatically with the launch of Apple's App Store and the ensuing boom in mobile videogame development. As one of the few commercial game engines optimized for iOS development available in the mid-2000s, and a relatively cheap resource for the typically small and independent teams seeking success on the mobile marketplace at the time,

---

[8] Incidentally, Unity Technologies ceased support for its 'Web Player' in 2015, after browsers such as Chrome began blocking Netscape Plugin Application Programming Interface (NPAPI) plugins. Many videogames developed for and published on Unity's Web Player have subsequently been lost.

Unity became something of an overnight success. Unity has since expanded to support multiplatform development, as well as a suite of other features including ad network and version control services. It can now be used to develop a wide variety of projects, and it supports development for an equally wide variety of platforms. In this way, Unity can be considered a general-purpose engine, meaning that although its default toolset is arguably geared towards the creation of particular types of content, it is built to accommodate a range of possible project types, design methodologies, and production workflows. Most game engines are, by contrast, built to accommodate very specific types of software development, and in this way, they have distinct affordances. A videogame studio might, for example, develop a custom engine that is specifically optimized for the development of first-person shooter videogames. Although Unity is a general-purpose engine, it still encourages developers to create content in specific ways. For example, Unity was initially optimized for 3D content creation, such that it was referred to as 'Unity3D' for many years. On Twitter, it is still common for developers to hashtag their Unity-related tweets with #Unity3D. Although it has since incorporated numerous tools and features for the creation of 2D content, Unity's underlying grain—its protocols, standards, and affordances—still enforces something of a 3D design methodology and epistemology. It is also important to note that Unity's affordances are relational,[9] in that they are shaped as much by the decisions of the company's key stakeholders and in-house software engineers as they are by the collective intelligence, ingenuity, and entrepreneurship of the engine's user base.

This relationality is especially evident in Unity's various 'platform' features. Unity's platform ecology will be explained in more detail in the following chapter, but for now, it is important to note that Unity encompasses not only a suite of software tools but also a (highly politicized) space of affective, cultural, and technological intermediation or 'platformization' (Nieborg and Poell 2018). This space encompasses, for example, the Unity Asset Store, where users can buy, sell, and/or freely obtain various user-developed assets and plugins. Unity also has something of a 'platform-based business model' (see Srnicek 2016; Nieborg and Poell 2018), which, in short, means that the majority of its users are

---

[9] See Bucher and Helmond (2017: 235) for a critical overview of affordances and relationality in connection with social media platforms.

offered a 'free' version of the core software in exchange for their data, their attention (broadly defined), their loyalty (cultivated through said attention), and their commitment to an EULA. Many Unity developers also identify as being part of a cultural scene that exists within the engine's platform ecology (Young 2018: 100–105). This scene includes like-minded developers, online forums, social events, and company 'evangelists',[10] each of whom share a common investment in the community, openness, and creativity that Unity seemingly affords. Although aspects of this scene may be considered antagonistic in relation to Unity's wider platform ecology, our sense is that the ecology itself—as a space of affective intermediation or, in Angela McRobbie's (2016) terms, a 'creativity *dispositif*'—explicitly works to contain what may have once been considered oppositional or countercultural videogame-making identities, attitudes, and practices.

Although Unity is considered a general-purpose engine, and although many developers stress that it can be customized in seemingly any direction, it is important to emphasize that, like all cultural software, Unity is not a neutral tool. From its default editing interface to its platform-based business model to its policy discourse of 'democratizing game development', Unity imposes a particular politics of software development. As one developer we interviewed for this book put it,

Yes, it's possible to work in many ways in Unity, but there is a way that they want you to work and you are pushed or encouraged to work within that environment, within those ways, right? And that's just the nature of any program, any software that you're using. Like, you write a document in Word and there are immediately decisions made for you about, like, what document it is, what an appropriate margin size is for a page, what a heading should look like, and things like that. And, even though you can modify those things, there's definitions there that are telling you what is acceptable and how to work within those spaces.

Of course, this is not to say that the shape and form of all cultural production today are entirely reducible to the cultural software that support it. For example, while it would be relatively uncontroversial to claim that Microsoft Word has a standardizing influence on word processing

---

[10] Unity's 'evangelist' representatives immerse themselves in regional communities and provide support for developers in the form of field engineering.

workflows, page layouts, and even spelling and grammar, it would be comparatively difficult (though not necessarily impossible) to prove that Microsoft Word has a standardizing influence on individual writing styles. It is nonetheless clear that word processing software and the process of writing are both grounded in what Matthew Kirschenbaum (2016: 7) calls an inseparable *materiality*, however difficult that materiality is to articulate (see Bogost [2006: 62] for a similar observation with regard to game engines). 'To know the software [of word processing]', writes Kirschenbaum (2016: 13), 'is to know something of the mind of the writer, however obliquely'. Similarly, while many of the developers we interviewed for this book firmly believed that Unity did not have a detectable influence on their creative output, they often (somewhat contradictorily) spoke of Unity imparting an ineffable 'look and feel', and expressed a keen interest in the various engines and tools used by their colleagues and peers. Like all cultural software, Unity *gives shape* to specific production workflows, design methodologies, software literacies, and modes of (self-)governance. These mediations cohere into a software circuit that utilizes multiple techniques—technological, discursive, and aesthetic—to draw users into its orbit. To paraphrase Kirschenbaum (2016: 7), most videogame developers would probably agree that their use of Unity is something that *should* be talked about, even if it is not immediately clear how or why. This book takes steps towards providing a critical vocabulary that, we hope, will encourage these conversations.

Book Outline and Context

This book draws on data from 175 semi-structured interviews with videogame developers, students, and educators, as well as participant observation and ethnographic fieldwork, conducted by both authors individually in Australia throughout 2018. Of our interviewees, 79% were men. While equal representation would be ideal, this percentage is reflective of the gender imbalance in the Australian videogame industry (GDAA 2018). The interviews were semi-structured and typically lasted between 30 and 60 minutes. Most (but not all) interviews were recorded, transcribed, and analysed for key themes. We conducted interviews with a wide range of developers, finding that the vast majority of respondents used Unity as their primary engine. We actively sought out interviews with users of, for example, Epic's Unreal engine—another key player in this space—though these users were in the minority.

Thus, while we occasionally draw on insights from interviews with Unreal developers and users of grassroots game engines such as Twine to make observations on game engines generally as a genre of cultural software, this book is primarily focused on Unity as its core case study. We also conducted ethnographic fieldwork and participant observation at a co-working space known as The Arcade, based in Melbourne's Southbank. Here, we were able to gain insight into the 'messy materiality' (Whitson 2018b) of videogame development, which was less apparent in semi-structured interviews. Most respondents agreed for us to use their real names in the research, though we have chosen to anonymize quotes where appropriate. As a tertiary source, we also draw from the experience of one author, Keogh, using Unity as a hobbyist videogame developer and videogame development educator.

Australia is a paradigmatic case study for understanding Unity's rise to prominence among student, independent, and professional videogame makers. In the early 2000s, before Unity 1.0 was released, many of Australia's most successful videogame studios were either acquired by or dependent on contracts from North American companies, in part because Australia had become something of a 'currency haven' in the wake of the Global Financial Crisis (GFC) (Banks and Cunningham 2016: 129; see also Darchen 2015: 209–210). During this time, Australian studios were turned into offshore subsidiaries and tasked with developing 'catalogue filler' software for North American publishers. As the effects of the GFC took hold in North America, many of these Australian subsidiaries were abandoned by their now-North American owners/investors, leading to a rapid deflation of the national industry and a diaspora of home-grown talent (see Apperley and Golding 2015; McCrea 2013). Out of this wreckage emerged a number of successful Australian mobile videogame developers and independent studios. Unity gained momentum in Australia for much the same reason that mobile videogame development gained momentum in Australia. That is, in the wake of the GFC, when a number of once-dominant studios went bankrupt, and when state support for media creatives dissipated, developers formed small teams and made 'small' videogames using the inexpensive and commercially feasible videogame-making resources available to them. According to a 2018 survey of 72 Australian videogame companies, 55 were using Unity, while seven were using Unreal, and five were using proprietary game engines (Game Developers' Association of Australia 2018). Internationally successful Australian mobile videogames such as *Crossy Road* (Hipster Whale, 2014) and, more recently, *Florence* (Mountains, 2018) were developed in Unity. Unity, together with Apple's App Store, has reinvigorated and reconfigured the Australian videogame industry over the past decade. At the same time, Unity's dominance in Australia has meant that a whole nation's videogame industry has, for several years, been more-or-less beholden to a single company's product. Australia is both an exceptional case study for understanding the mediations of Unity and a hugely relevant one as other regional videogame industries are also emerging as somewhat autonomous from the traditionally dominant blockbuster studios concentrated in North America and Japan. The global videogame industry is intensely local (Kerr 2017), and concentrating our case study on a single local industry (albeit in a trans-local context) grounds our analysis.

We proceed in Chapter 2 by setting up the story of Unity—its socio-historical context, its political economy, and its platform-based business model—before considering its articulations within and across the circuits of cultural software and the field of videogame production more broadly. In Chapter 3, we first consider the various workflows that have become articulated to Unity. In Chapter 4, we then look at the engine's underlying grain, which has a direct bearing on the design methodologies adopted by its users and, by extension, the cultural reception of Unity-developed content. Chapter 5 then explores the various literacies that have emerged within and around Unity, while Chapter 6 concludes by considering the governance structures associated with the engine's democratization *dispositif*—that is, the mechanisms and knowledges through which potential users are enrolled into the engine's software ecology. A common finding across each of the circuits is that Unity is experienced as simultaneously empowering and disempowering. Unity provides accessible design standards for videogame developers but, in doing so, reveals a lack of accessible design principles. Unity is viewed as creatively enabling for non-programmers, yet many speak of the technical know-how necessary to deviate from its 'path of least resistance'. Unity extracts surplus value from the informal support labour of its asset store developers, yet it utilizes the revenue obtained from its more formalized actor-networks to provide support for its informal user base. Unity's open licensing structure makes it 'easier' for students and independents to create content, yet it also leads to anxieties around monopolization and homogenization. Unity demystifies the process of development and makes visible a wider diversity of creators and genres, yet it also conceals the technical and economic infrastructures that underlie its user-friendly interface. This brings us back to the duality identified in the opening anecdote: How might we make sense of a game engine that, according to popular discourses, both democratizes and homogenizes, that generates enthusiasm and anxiety in equal measure?

References

Adorno, Theodor, and Horkheimer, Max. 2002 (1947). *Dialectic of Enlightenment: Philosophical Fragments*. Stanford: Stanford University Press.

Apperley, Thomas H., and Daniel Golding. 2015. "Australia." In *Video Games Around the World*, edited by Mark J. P. Wolf and Toru Iwatani. Cambridge: MIT Press.

Apperely, Thomas H., Darshana Jayemanne, and Bjorn Nansen. 2016. "Postdigital Literacies: Materiality, Mobility and the Aesthetics of Recruitment." In *Literacy, Media and Technology: Past, Present and Future*, edited by Becky Parry, Cathy Burnett, and Guy Merchant, 203–218. New York: Bloomsbury.

Ash, James. 2015. *The Interface Envelope: Gaming, Technology, Power*. New York: Bloomsbury.

Banks, John. 2013. *Co-creating Videogames*. New York: Bloomsbury.

Banks, John, and Stuart Cunningham. 2016. "Creative Destruction in the Australian Videogames Industry." *Media International Australia* 160 (1): 127–139.

Berry, David M., and Michael Dieter, eds. 2015. *Postdigital Aesthetics: Art, Computation and Design*. London: Palgrave Macmillan.

Bogost, Ian. 2006. *Unit Operations: An Approach to Videogame Criticism*. Cambridge: MIT Press.

Bratton, Benjamin H. 2015. *The Stack: On Software and Sovereignty*. Cambridge: MIT Press.

Brodkin, Jon. 2013. "How Unity3D Became a Game-Development Beast." Dice, June 3. https://insights.dice.com/2013/06/03/how-unity3d-become-a-game-development-beast/.

Bucher, Taina, and Anne Helmond. 2017. "The Affordances of Social Media Platforms." In *SAGE Handbook of Social Media*, edited by Jean Burgess, Thomas Poell, and Alice Marwick, 234–253. Los Angeles: SAGE Publications.

Chun, Wendy Hui Kyong. 2011. *Programmed Visions: Software and Memory*. Cambridge: MIT Press.

Darchen, Sébastien. 2015. "'Clusters' or 'Communities'? Analysing the Spatial Agglomeration of Video Game Companies in Australia." *Urban Geography* 23 (2): 202–222.

Dillet, Romain. 2018. "Unity CEO Says Half of All Games Are Built on Unity." Techcrunch, September 5. https://techcrunch.com/2018/09/05/unity-ceo-says-half-of-all-games-are-built-on-unity/.

du Gay, Paul, Stuart Hall, Linda Janes, Hugh Mackay, and Keight Negus. 1997. *Doing Cultural Studies: The Story of the Sony Walkman.* London: Thousand Oaks.

Foucault, Michel. 2008. *The Birth of Biopolitics: Lectures at the Collège de France, 1978–79.* Basingstoke: Palgrave Macmillan.

Freedman, Eric. 2018a. "Engineering Queerness in the Game Development Pipeline." *Game Studies* 18 (3). http://gamestudies.org/1803/articles/ericfreedman.

Freedman, Eric. 2018b. "Software." In *The Craft of Criticism: Critical Media Studies in Practice,* edited by Michael Kackman and Mary Celeste Kearney, 318–330. New York: Routledge.

Game Developers' Association of Australia. 2018. "What Game Engines Are Companies Using?" https://www.facebook.com/GameDevAssocAus/photos/a.1943622552593569/2062766267345863/?type=3&theater.

Grayson, Nathan. 2018. "No, *PUBG* Is Not an 'Asset Flip'." Kotaku, 18 June. https://kotaku.com/no-pubg-is-not-an-asset-flip-1826935848.

Grossberg, Lawrence. 1995. "Cultural Studies: What's in a Name (One More Time)." *Taboo: The Journal of Culture and Education* 1: 1–37.

Grossberg, Lawrence. 1997. *Bringing It All Back Home: Essays on Cultural Studies.* Durham: Durham University Press.

Grossberg, Lawrence. 2010. *Cultural Studies in the Future Tense.* Durham: Duke University Press.

Haas, John. 2014. "A History of the Unity Game Engine." Interactive Qualifying Project. Worcester Polytechnic Institute. https://web.wpi.edu/Pubs/E-project/Available/E-project-030614-143124/.

Harvey, Alison. 2014. "Twine' Revolution: Democratization, Depoliticization, and the Queering of Game Design." *Game* 3: 95–107.

Jenkins, Henry. 2006. *Convergence Culture: When Old and New Media Collide.* New York: New York University Press.

Keogh, Brendan. 2019. "From Aggressively Formalised to Intensely In/Formalised: Accounting for a Wider Range of Videogame Development Practices." *Creative Industries Journal* 12 (1): 14–33.

Kerr, Aphra. 2017. *Global Games: Production, Circulation and Policy in the Networked Era.* Routledge: New York.

Kirkpatrick, Graeme. 2013. *Computer Games and the Social Imaginary.* Cambridge: Polity Press.

Kirschenbaum, Matthew G. 2016. *Track Changes: A Literary History of Word Processing*. Cambridge, MA: Harvard University Press.

Kline, Stephen, Dyer-Witheford, Nick, and de Peuter, Greig. 2003. *Digital Play: The Interaction of Technology, Culture and Marketing*. Montréal: McGill-Queen's University Press.

Lévy, Pierre. 1997. *Collective Intelligence: Mankind's Emerging World in Cyberspace*. Cambridge: Perseus Books.

Lobato, Ramon, and Julian Thomas. 2015. *The Informal Media Economy*. Hoboken, NJ: Wiley.

Manovich, Lev. 2013. *Software Takes Command*. Cambridge: MIT Press.

McCrea, Christian. 2013. "Australian Video Games: The Collapse and Reconstruction of an Industry." *Gaming Globally*, edited by Nina B. Huntsman and Ben Aslinger, 203–207. New York: Palgrave Macmillan.

McRobbie, Angela. 2016. *Be Creative: Making a Living in the New Culture Industries*. Cambridge: Polity Press.

Nieborg, David B., and Shenja van der Graaf. 2008. "The Mod Industries? The Industrial Logic of Non-market Game Production." *European Journal of Cultural Studies* 11 (2): 177–195.

Nieborg, David B., and Thomas Poell. 2018. "The Platformization of Cultural Production: Theorizing the Contingent Cultural Commodity." *New Media & Society* 20 (11): 4275–4292.

O'Donnell, Casey. 2014. *Developer's Dilemma*. Cambridge: MIT Press.

Pedercini, Paolo. 2017. "Indiepocalypse Now: MadMaxing Attention Economies in the Age of Cultural Overproduction." Molleindustria. http://molleindustria.org/indiepocalypse/.

Qiu, Jack Linchuan, Melissa Gregg, and Kate Crawford. 2014. "Circuits of Labour: A Labour Theory of the iPhone Era." *triple* 12 (2): 564–581.

Srnicek, Nick. 2016. *Platform Capitalism*. Malden: Polity Press.

Unity. 2018. "Unity at GDC Keynote—March 19, 2018." YouTube, March 19. https://www.youtube.com/watch?v=cmRSkHl-Gv0.

Whitson, Jennifer R. 2018a. "Voodoo Software and Boundary Objects in Game Development: How Developers Collaborate and Conflict with Game Engines and Art Tools." *New Media & Society* 20 (7): 2315–2332.

Whitson, Jennifer R. 2018b. "What Can We Learn From Studio Studies Ethnographies? A 'Messy' Account of Game Development, Materiality, Learning, and Expertise." *Games and Culture* (OnlineFirst). https://doi.org/10.1177/1555412018783320.

Young, Christopher J. 2018. "Game Changers: Everyday Gamemakers and the Development of the Video Game Industry." PhD diss., University of Toronto.

Zhao, Sumin, Emilia Djonov, and Theo van Leeuwen. 2014. "Semiotic Technology and Practice: A Multimodal Social Semiotic Approach to PowerPoint." *Text&Talk* 34 (3): 349–375.

CHAPTER 2

Unity's Socio-historical Context and Political Economy

Abstract This chapter argues that Unity's 'conditions of existence' are predicated on a long history of developer- and player-oriented videogame-making tools, practices, and communities, and that the engine's business model is consonant with a broader 'platformization of cultural production' in today's media industries. It describes the emergence of proprietary game engines in the early 1990s in terms of a broader shift from programmer-centric development to content-centric development. It argues that Unity builds on long-standing agitations for 'democratized' tools in videogame development, such as those associated with modding scenes and indie development. It then discusses Unity's platform-based business model, touching on the engine's licensing structure; its revenue model; its asset store; and its attempt to establish spaces of 'affective intermediation' in videogame culture.

Keywords Videogame history · History of game engines ·
Unity game engine · Videogame development · Asset store ·
Platformization of cultural production

Modes of production in the videogame industry have undergone fundamental changes over the past decade. Throughout the 1990s and early 2000s, videogame studios would typically create bespoke development tools from scratch, for use on internal projects. Developers would need to secure publishing deals and software development kits

© The Author(s) 2019   23
B. Nicoll and B. Keogh, *The Unity Game Engine and the Circuits of Cultural Software*, https://doi.org/10.1007/978-3-030-25012-6_2

(SDKs)[1] from publishers and other brokerage actors if they were to have any hope of releasing their content on commercial videogame hardware (see Kerr 2017). With a few notable exceptions, companies did not share their tools (unless leasing for a fee) or openly trade industry knowledge (O'Donnell 2014; see also Darchen 2015: 213–214 for an Australian perspective). For the most part, development was slow, secretive, resource-intensive, and open only to those with access—financially and institutionally—to the proprietary pipelines, toolchains, and skill sets required to produce and publish videogame content.

Since the early 2000s, however, several low-cost and low barrier to entry game engines have enabled a wider diversity of creators to make videogames. Unity is central to this transformation, but it did not emerge in a vacuum. Its existence is indebted to a long history of developer- and player-oriented videogame-making tools, practices, and communities, and its business model is consonant with a broader 'platformization of cultural production' in today's media industries (see Nieborg and Poell 2018). In this chapter, we set the scene for Unity's 'conditions of existence' (Parikka 2012: 6) by looking at the various historical circumstances that contributed to its emergence, as well as the political economy[2] of its business model and licensing structure. Along the way, we look at the history of videogame-making tools, both digital and non-digital; the shift from programmer-centric development to content-centric development, which occurred in the 1990s; the concurrent rise of proprietary game engines; the long history of grassroots and countercultural videogame-making practices; and the rise of Unity as a platform-based software ecology for videogame production.

---

[1] SDKs are software resources that hardware manufacturers and publishers lease to videogame developers in the final stages of a project. Companies such as Nintendo, Sony, and Apple build protocols into their platforms that render it impossible to run a game on their hardware without the requisite SDK (see O'Donnell 2014: 200–207).

[2] The political economic tradition in media studies is broadly focused on the following questions: Who owns power in the media industries; how is that power consolidated and filtered through an organization's supply chains, conditions of labour, and revenue models; and how do these power structures fall upon media audiences, consumers, and users?

VIDEOGAME DEVELOPMENT BEFORE GAME ENGINES

Game engines and other kinds of software tools have not always played such a prominent role in videogame development. Early videogames such as *Tennis for Two* (Higginbotham, 1958) were hardwired directly onto vacuum tube analogue computers and synced to oscilloscope displays. Likewise, early arcade videogames such as *Computer Space* (Syzygy Engineering, 1971) ran on hardwired integrated circuits rather than read-only memory (ROM) storage devices. Many early videogame consoles and computers did not have code frameworks upon which content could be built; instead, programmers created content from first principles. The Atari Video Computer System, for example, did not have an operating system, meaning that programmers were 'responsible for handling every interaction on the machine'—even the console's on-off and reset switches, whose functions had to be written into each individual cartridge program (Montfort and Bogost 2009: 34).

In the 1980s, various consumer-grade videogame-making tools were developed for microcomputers such as the Apple II, the Commodore 64, and the Atari 8-bit family. An oft-cited example is *Pinball Construction Set* (*PCS*; BudgeCo, 1983), developed by Bill Budge and published by Electronic Arts for the Apple II in 1983 (see Barton and Loguidice 2009). *PCS* consists of a drag-and-drop inventory of digital pinball parts—flippers, bumpers, spinners, and so on—that can be assembled into playable pinball videogames. *PCS* also enabled users to customize art, music, and physics for their digital pinball creations. A completed *PCS* build can be published on floppy disk and played on another Apple II machine, regardless of whether that machine has the core *PCS* software installed. *PCS* spawned a whole series of popular 'construction set' videogames, including *Will Harvey's Music Construction Set* (Harvey, 1984), *Racing Destruction Set* (Koenig, 1985), and *Adventure Construction Set* (Smith, 1984), the latter of which introduced a basic scripting tool. The proliferation of—and widespread interest in—videogame-making tools in the 1980s was precipitated by a broader trend of hobbyist programming on microcomputers. Regional microcomputer scenes—wherein hobbyists wrote and exchanged their own software programs—flourished across Europe, South America, and Australasia in the 1980s. In these contexts, magazines such as *Compute!'s Gazette* (*sic*) functioned as videogame-making 'tools' of sorts, to the extent that they published 'type-in programs' (often written in BASIC) that readers could manually enter, line by line, on their home machines. During this era, it was also common for developers to sketch design concepts and graphics on paper before translating said concepts and graphics into code (Stuckey et al. 2015: n.p.). There was, in essence, a much smaller division between those who played videogames and those who made them, and a much more 'informal' landscape of videogame development as a result (Keogh 2019).

It would also be remiss for us not to mention the long history of non-digital game-making tools, practices, and communities. For example, tabletop 'war games' of the 1970s often combined rule-based systems with player-oriented scenario editors. Like videogame engines, these 'war engines' (Lowood 2016a) emerged from a need to streamline the increasingly content-heavy nature of (non-digital) game development. War engines enabled players to generate their own combat scenarios rather than simply play scenarios created by game designers. While Henry Lowood (2016a: 103) suggests that there 'are indeed meaningful similarities' between war engines and videogame engines, he stresses that 'this interpretative strategy might be a stretch', insofar as the war engine's underlying systems are transparent and modular, while the videogame engine's 'secrets' tend to be black-boxed or closed source (Lowood 2016a: 103). There are nonetheless important continuities between tabletop game engines and videogame engines. Namely, Lowood (2016a: 101–102) identifies something of an enduring 'imaginary' in game development that has persisted since the era of war engines: the fantasy of a game-making tool that, in the vein of a universal Turing machine (or, perhaps more accurately, in the popular and arguably incorrect reception of the idea of a universal Turing machine [see Sack 2019: 54–55]), can do anything and everything (cf. Banks 2013: 51–53).[3] As we will see in the following chapters, Unity's generalist nature places it squarely within this imaginary.

Videogame engines were 'driven into existence' (Lowood 2016a: 103) for similar reasons to tabletop engines. As videogames became increasingly content-heavy in the 1990s—that is as videogame development became increasingly dependent on the labour of designers and artists rather than programmers—studios began to create tools for the purposes of coordinating (and patenting) the various part(icipant)s

---

[3] In the context of war games, this imaginary expressed itself in the desire for a war engine 'with unlimited scenario generation powers' (Lowood 2016a: 102).

involved in the development pipeline. This history constitutes what we call a developer-oriented lineage of proprietary and commercial game engines.

DEVELOPER-ORIENTED GAME ENGINES: FROM PROPRIETARY TO COMMERCIAL

The term 'game engine' first emerged in 1993 to describe a specific class of software objects (Lowood 2016b: 203; Bogost 2006: 60). As mentioned above, game engines emerged partly in response to the increasingly content-heavy nature of professional videogame development in the mid- to late 1990s. 'Content' here refers to a videogame's representational elements—assets, animations, audio, narrative, writing, level design, and so on—rather than its code, systems, and procedures. Prior to the emergence of game engines, programmers and software engineers were understood to be the most central—if not exclusive—labourers associated with videogame development; they wrote code, designed systems, and created (often rudimentary) art and music.[4] In the 1990s, the proliferation of dedicated videogame hardware and a growing consumer appetite for 'good graphics' and intellectually stimulating 'gameplay' (Kirkpatrick 2015: 60–61) meant that programmers were spending more time on narrative, design, and art than they had in the past. One of our respondents described this shift in terms of a transition from programmer-centric development to content-centric development (cf. Freedman 2018: n.p.). In content-centric development, the bulk of development is focused on design and art rather than programming. It therefore requires a whole composition of different developers with different skill sets—not only programmers and software engineers but also designers, artists, writers, composers, producers, and other actors—to collaborate and contribute to the development process.

---

[4]This historical generalization risks erasing other types of labour associated with videogame development. We could, for example, consider the labour of Roberta Williams, who, in 1979, designed the graphic adventure videogame *Mystery House* (Williams, 1980). According to Laine Nooney (2013: n.p.), Williams 'was not a programmer; she was a housewife and mother of two', and her 'design contained no code, no instruction sets, no sense of how the game she wrote would function on a computer'. Nooney makes the important point that videogame history struggles to 'make sense' of figures such as Roberta Williams for this reason.

In content-centric development, programmers are no longer responsible for art and design, but rather for creating the tools that designers and artists can use to author and edit content.[5] Content-centric development therefore requires pipelines that make it easier to manage each part(icipant) in the development process. Pipelines consist of company-specific development tools and techniques through which content is not only created but also continually iterated upon. Being able to iterate on content that has already been created is made easier with tools that are themselves iterative—that is, tools that can be customized to suit a variety of different (and constantly evolving) videogame projects and disciplinary roles. Given that programmers were (and still are, in many studios) responsible for creating and maintaining tool pipelines, it is often the case that these tools can implicitly delegate power to programmers and software engineers rather than designers and artists (see Whitson 2018a).

In the 1990s and 2000s, many videogame studios created proprietary toolsets (that is, game engines) to streamline and patent their in-house content-centric development processes. These proprietary engines were not singular objects but rather amorphous collections (or 'libraries') of tools that could be uniquely tailored to company-specific projects, workflows, and player-oriented 'modding' practices (see Banks 2013: 45). In Eric Freedman's (2018: n.p.) words, proprietary engines were developed not only to 'realize greater efficiencies in the game development pipeline (marrying code and design)' but, crucially, 'to mitigate the [company's] dependency on the middleware of other software companies'. The prototypical example here is the Doom engine (also known as id Tech 1), created by North American company id Software in 1993 (see Lowood 2016b). id Software used the Doom engine to develop *Doom* (id Software, 1993) and its sequel. Sometimes, companies such as id Software would license their proprietary engines to other companies or make them partially accessible for public use. Content produced or published using the Doom engine would, in these instances, be considered a 'proprietary extension' of id Software's patented engine technology

---

[5]Eric Freedman (2018: n.p.) goes so far as to argue that '[t]his industrial division also shaped the field of game studies, placing more focused attention on visual analysis, ignoring certain material relations to study narrative, genre, seriality and other literary and cinematic conceits'.

(see Nieborg and van der Graaf 2008).[6] Shortly after they were introduced in studio environments, game engines entered the public imagination. Videogames were often promoted on the basis that they were
'powered by' particular engines—engines that players, critics, and magazines came to recognize and associate with, for example, good graphics
and stable frame rates (see Chapter 5). Moreover, by generating enthusiasm for their engines, companies such as id Software were able to advertise their tools to prospective licensees.

The patenting (and occasional leasing) of company-specific game
engines throughout the 1990s and early 2000s was one among several
factors that contributed to a culture of protectionism and secrecy among
videogame developers (O'Donnell 2014). Studios would need to hire
multiple software engineers to create, maintain, and iterate upon their
tools, often over a period of several years. As one of our respondents
put it, 'code goes stale', even when written in-house. Software tools—
regardless of whether they are custom-made within studios or commercially available for public use—need to be constantly updated to comply
with the latest hardware and software standards. Unity, for example, is a
product of the labour of thousands of software engineers and everyday
users, who continuously update, patch, and contribute resources to the
engine. A comparable amount of labour, capital, and resources would
be required to build a custom engine: personnel to create and maintain
the engine, time (often years) to develop the engine itself, as well as an
ongoing allocation of resources to engine upkeep. This is one reason
(among many) why proprietary engines have historically been treated
as intellectual property and safeguarded within studios. This safeguard-
ing of industry tools created a culture of secrecy and protectionism
that, according to Casey O'Donnell (2014: 273), was 'both *top-down*—
non-disclosure agreements, closed licensing structures, proprietary hardware and software—and *bottom-up*—"my idea is super secret and super
awesome," [...] "if I talk about this, someone is going to steal it"'.

Although some of our respondents reminisced positively about their
work with and on proprietary engines, most associated them with an
industry environment that was detrimental to interdisciplinary collaboration. One software engineer, who had worked as an engine developer

---

[6]id Software tend to open source their engines after several years, meaning that content
produced on their engines after the transition to open source may no longer be considered
a proprietary extension.

at several Australian videogame and software companies throughout the 2000s, remarked that he 'cherished' his time working on proprietary engines, because he was able to acquire a foundational knowledge of various engine technologies. Yet, even he acknowledged that most people involved in videogame development were put at a significant disadvantage by proprietary engines because, as he put it, 'a lot of people had that experience where there was a mysterious engine maker that just said, "this is how it is, this is what you've got, deal with it"'. The situation was even more challenging for developers who did not work in studio spaces. Independent developers, for example, would need to create tools from scratch or rely on restricted public access to proprietary engines. Freelance programmers and artists would need to learn and relearn company-specific tools as they moved between jobs. Students learned design skills by using the player-oriented level editors and modding toolkits available to them, rather than the tools actually used in industry. In his 2014 book *Developer's Dilemma*, O'Donnell (2014) suggests that developers can alleviate these problems by embracing the open sharing of design standards, tools, and practices—to, in a word, 'democratize' videogame development. Unity comes close to facilitating such a movement, but it does so on the basis of its quasi-monopolistic status.

Opaque publishing procedures, hierarchical organizational structures, and precarious forms of employment remain ever-present in many (though not all)[7] videogame studios, yet commercial game engines such as Unity have arguably simplified and demystified the process of developing multiplatform videogame content. Where once videogame development was characterized by 'waterfall' management techniques and hierarchical production pipelines, commercial game engines have contributed to what Whitson (2018b: 9) calls an 'increasingly heterogeneous' landscape of development: '[t]eams are often smaller, organized in a "flat" rather than hierarchical manner, and multitask rather than occupy the discrete roles depicted in textbook organizational charts'. This corresponds to a shift from what some developers call 'waterfall' to 'agile' management processes. In waterfall management, the various steps involved in a development process are itemized in design documents, assigned to various team members, and translated into milestones.

---

[7] Kristine Jørgensen (2017: 15), for example, discusses a Norwegian studio that adopts a decentralized mode of governance and explicitly 'restricts extended use of crunch time and other unethical activities'.

Agile management is, according to Kristine Jørgensen (2017: 11), a more 'flexible and dynamic method' by contrast, one 'that focuses on iterative work with frequent builds, thus stressing the importance of having functional versions of the software over detailed documentation of planned functionality'. As will be discussed in Chapter 4, rapid iteration and prototyping are cornerstones of Unity's editing interface.

Today, most developers opt to use commercial engines such as Unity or Unreal as opposed to building an engine from scratch or obtaining a licence to use a proprietary engine. However, there still exists an incentive for multinational videogame companies such as Rockstar and Electronic Arts to develop proprietary engines. In our research, we heard anecdotes about how 'triple-a' or blockbuster studios often use commercial engines such as Unity to prototype ideas, before switching to proprietary tools for the final project. Triple-a studios tend to have the budgets to justify developing custom tools, and it makes legal and economic sense for them to invest in internal research and development rather than renting services and technologies from competitors. Furthermore, triple-a companies are often creating 'high performance' videogames with specialized content requirements—they are working at the 'bleeding edge' of software development—meaning that general-purpose engines may not be considered performant enough for their purposes. The challenge for a commercial game engine such as Unity is therefore to strike a balance between accessibility and performativity. Commercial engines need to be general-purpose enough to attract developers of all skill levels and disciplinary backgrounds, yet also performant enough to attract medium-sized studios, developers of mobile videogames, and perhaps even triple-a companies, who promise significant returns on royalties as well as increased visibility in culture.

PLAYER-ORIENTED GAME ENGINES AND GRASSROOTS VIDEOGAME-MAKING PRACTICES

Just as game engines have an important lineage in the history of developer-oriented production processes, they have an equally important (and oftentimes overlapping) lineage in the history of grassroots videogame-making tools, practices, and communities. Grassroots videogame-making practices have always existed alongside—and are even synonymous with—the development of videogames as a medium.

Early videogames such as *Tennis for Two* and *Spacewar!* (Russel, 1962) were products of a 'hacker' culture that, while gendered and exclusionary, established the idea that computers could be used for playful, experimental, and even countercultural purposes (Turkle 2005; see also Dyer-Witheford and de Peuter 2009: 10). *Tennis for Two* and *Spacewar!* were effectively 'modifications' of military technologies explicitly developed for purposes related to, for example, nuclear physics. Likewise, computer games developed by hobbyists in the 1980s were, according to Graeme Kirkpatrick (2013: 65), largely responsible for 'naturalizing' early home computer interfaces—interfaces that were not optimized for playing videogames but rather for running (oftentimes deeply unintuitive) productivity applications (see also Swalwell 2012). In many regions, computer and console videogames were pirated, cloned, or creatively remixed for local audiences (see Gazzard 2014; Švelch 2018; Nicoll 2019), long before the widespread adoption of modding toolkits. Videogames have always been subject to (even shaped by) regional, transnational, and local articulations of player-oriented consumption and appropriation, including 'paratextual' activities such as fan art and unofficial translations (Ng 2009). As a medium, videogames are a product of an ongoing exchange between informal and formal processes of production and consumption, including hacking, pirating, remixing, and hobbyist programming.

Modding is an important and widely discussed player-oriented videogame-making practice with a long history, albeit one that the industry has frequently sought to co-opt as a desirable commercial activity. Modding usually refers to the process of using tools to modify an existing videogame by, for example, adding characters, content, or features not present in the original. Modding can be an informal practice not envisioned by developers, or a practice actively encouraged by developers through the availability of player-oriented level editors and toolkits. A famous example is *Counter-Strike* (Valve, 1999), a 1999 modification of Valve's first-person shooter, *Half-Life* (Valve, 1998). Both *Half-Life* and *Counter-Strike* were built using Valve's Goldsource engine, which was itself a heavily modified version of id Software's open-source Quake engine. Valve released Goldsource's level editor for public use, which led to the emergence of many fan-made Half-Life 'mods', the most successful of which was *Counter-Strike*. *Counter-Strike* was so popular that its intellectual property was (re)acquired by Valve—as were the modders who created it—and published as an official Valve game in 2000.

The absorption of modding practices into commercial videogame-making is not a fringe phenomenon, but a significant aspect of the videogame industry. Many of the most commercially successful videogame franchises and genres from the past decade began their lives as mods. The multiplayer online battle arena (MOBA) genre, for example, began life as a player-made mod of Blizzard's *Warcraft III: Reign of Chaos* (Blizzard, 2002). The battle royale genre, which exploded in popularity alongside *Fortnite* (Epic Games, 2017), originated with PlayerUnknown's mod of *ARMA 3* (Bohemia Interactive, 2013) and the subsequent stand-alone *PlayerUnknown's Battleground* (PUBG Corporation, 2017). It is worth emphasizing, however, that most player-oriented level editors do not provide access to an engine's underlying source code or even an engine's full suite of tools. In fact, it is relatively uncommon for developers to open any part of their engines for public consumption (see van der Graaf 2018: 14). Modding toolkits were nevertheless the primary means through which hobbyists could access otherwise-proprietary videogame-making resources in the 1990s and 2000s.

Player-oriented videogame-making practices grew in popularity alongside a broader enthusiasm for what Henry Jenkins (2006) calls 'cultural convergence' in the 2000s. In the wake of the dot-com crash in the early 2000s, many media organizations turned to business models that embraced user participation, produsage, and the 'platformization of cultural production' (Nieborg and Poell 2018). For Jenkins, this shift was associated with a blurring of producer and consumer roles, such that everyday users were more empowered to create, edit, share, and remix content. In Jenkins's (2006: 18) account, this process was fundamentally democratic, because it implied that everyday users could wrest power from media companies and 'bring the flow of media more fully under their control'. However, since that time, cultural convergence has been assimilated into the underlying production logics of platform companies, who actively encourage user participation as part of their core business models. This is especially evident in the videogame industry where, since the early 2000s, companies such as Sony and Nintendo have actively promoted user participation through 'flagship' videogames such as *LittleBigPlanet* (Media Molecule, 2008) and *Super Mario Maker 2* (Nintendo, 2019) (see Sotamaa 2010).

These shifts were also concurrent with the emergence and widespread adoption of commercial videogame-making tools in the 2000s and early 2010s. Adobe's Flash, for example, was not explicitly created with videogame development in mind, but was nevertheless adopted by independents, hobbyists, and students as a videogame-making tool. Tools such as Flash, GameMaker, RPG Maker, and Twine supported (and continue to support) vastly different developer communities, but they can be identified as part of a broader movement in and around the 2000s towards more accessible and equitable ways of making and playing videogames. Throughout this period, a number of fringe developers and communities aggravated for change in the videogame industry through blog posts, community events, manifestos, and books (see, e.g., Anthropy 2012). The 'indie' movement, which also gained traction in the late 2000s, carried a similar—if somewhat less countercultural—promise. The indie ethos is notoriously difficult to pin down (see Lipkin 2013), but the general idea is to develop videogames on one's own creative and financial terms, rather than those dictated by studios, publishers, and hardware manufacturers. Many indie developers opt for a life of precariousness and uncertainty in pursuit of the 'dream' of creative and financial independence.

Each of the above historical developments—modding, cultural convergence and the platformization of cultural production, the indie ethos, and the ongoing fight for low-cost and low barrier to entry videogame-making tools and resources—created an 'opportunity space' for Unity to emerge and establish an identity for itself. Unity capitalizes on the countercultural spirit underlying these various movements and strategically positions itself at the intersection of corporate, grassroots, formal, and informal videogame-making practices. It can be viewed as an appeal to long-standing fantasies of a game engine that can 'do anything and everything', and a response to the history, culture, and mythos of grassroots videogame-making practices. In 2014, John Riccitiello, ex-CEO of Electronic Arts, took over from David Helgason as the CEO of Unity Technologies. In a blog post, Helgason (2014: n.p.), Unity Technologies's former CEO, wrote that Riccitiello is 'the right person to help guide the company to the mission that we set out for ourselves over a decade ago: democratize game development!'. Although it is doubtful that Unity Technologies's mission statement has always been to democratize videogame development, it is clear that the company has since adopted a 'discourse of democratization' as its guiding philosophy, and that this discourse has discovered profound resonances across formal, informal, and interregional videogame-making communities.

However, if Unity has contributed to a more democratized landscape of videogame development—already a contentious and complicated claim, and one that will be explored in more detail in Chapter 6—then it has done so not solely on its own laurels, but rather by building on a long history of grassroots videogame-making practices, tools, and communities. Moreover, Unity users have played an important role in developing the engine into what it is today. Users have, for example, developed thousands of assets and plugins for the Unity Asset Store, many of which have since been absorbed into the engine's default toolset. Users have also volunteered knowledge, advice, and tutorials about Unity on community forums and blogs—knowledge that, moreover, was lacking in official documentation in the engine's early years (Haas 2014: 23)—creating a pool of collective intelligence that our respondents often linked to the engine's supposedly democratizing effects. Likewise, users have shared data with Unity Technologies—knowingly or otherwise—which has enabled the company to rapidly scale up and optimize its services. Media historians have long noted that narratives of technological development are often conceptualized through what Lisa Gitelman (2006: 61) terms a 'production/consumption dichotomy' that, as in Unity's origin-story, place a heavy emphasis on the (very often male) figures responsible for 'inventing' and 'pioneering' the technology in question, while sidelining (and coding as feminine) narratives of media consumption, practice, and appropriation. An argument developed in Chapter 6 is that Unity users do not give themselves enough credit when it comes to their role in making the tools of videogame development more accessible and, indeed, democratic.

UNITY'S PLATFORM ECOLOGY

One of Unity's most defining features is its 'platform-based' business model. According to Riccitiello (in Takahashi 2018: n.p.), Unity is 'responsible for more than half' of all videogames published on commercial platforms and, as of September 2018, is currently being used by studios in every country except the Vatican, South Sudan, and North Korea. In other words, Unity has established powerful 'network effects' in the global videogame-making ecology—it has territorialized interregional videogame communities, interpellating and erasing regionally specific videogame-making identities and practices in the process (Vogel 2017)—and it has done so largely on the basis of its platform-based business model. In *Platform Capitalism*, Nick Srnicek (2016) observes that platform companies such as Google and Facebook seek to generate network effects by creating powerful incentives for users to sign up to their platforms. They often do this by providing an (ostensibly) free service or tool, such as a social media account or an application program interface (API), as well as by enabling various actors (users, advertisers, bots, developers, and so on) to interact on the platform in ways that are (ostensibly) mutually beneficial. Platforms cross-subsidize the costs associated with maintaining these free services and tools by raising prices on other elements of their businesses (such as advertising) or by soliciting welfare support from venture capital investors. Once in a monopoly position, platforms can then expand their data collection on users—data that, once refined, can be used to generate further revenue and attract further investment. The platform economy is thus characterized by what David Nieborg and Thomas Poell (2018: 4282) call 'strong *winner-takes-all* effects' or monopoly tendencies. Once a particular platform establishes network effects in its area—whether that be social media, transport, e-commerce, or cultural production—it becomes very difficult to dislodge that platform. Many of the terms used to describe platforms—such as infrastructures, intermediaries, openness, and, indeed, the platform metaphor itself—are decidedly value-neutral, and they obscure the fact that platforms are political entities that regularly flout their ethical, social, and economic responsibilities (Gillespie 2017; Tkacz 2014).[8]

Unity makes use of multiple platform-based business techniques, the first of which is its licensing structure. Unity Technologies offers multiple licences for its software. The 'Personal' licence provides 'free' access to the core engine software, while the 'Plus' and 'Pro' versions (USD$25/month and USD$125/month, respectively, at the time of writing) provide additional services, such as access to technical support and analytics. It is important to note, however, that the Personal version of Unity is only free insofar as no money is exchanged when users first download it. Users pay for Unity with their data, their commitment to an end-user

---

[8] As this book was going to print, news broke of a sexual harassment lawsuit filed against Riccitiello and other Unity Technologies management by one of the company's former senior directors. The complainant 'detailed that her time with the company was fraught with inappropriate comments from male management towards women' (Lanier 2019). This instance provides a crucial reminder that so-called platforms are not detached from the broader social and political issues at stake in the software and videogame industries.

licence agreement (EULA), and their contribution to the engine's network effects. Most of our respondents—including medium-sized studios—used the Personal version of Unity. At present, the Personal licence has three key stipulations, which are enforced through the EULA. The first stipulation is that a percentage of royalties on Unity-developed projects are payable to Unity Technologies if the project surpasses an annual revenue threshold of USD$100,000. However, very few projects developed under the Personal licence ever surpass this threshold. To this end, the second key stipulation is that all users must agree to share data with Unity. Unity collects data within its actual editing interface (so that it can perform analytics on its own software), as well as data on where and how Unity-developed products are distributed and purchased once published (so that the company can gain insight into broader market dynamics). The third key stipulation—unique to the Personal licence—is that a Unity 'splash screen' must feature in the opening credits of a Personal licence-developed product. The Plus licence increases the royalties threshold to USD$200,000, while the Pro licence places no limits on annual earnings. There is also a separate licence for developers using Unity to create and publish gambling software.[9] Each of the above stipulations is subject to change at any moment and probably will change by the time this book is published.

None of our respondents—even those working on potentially lucrative Unity-developed projects—had been directly approached by Unity Technologies regarding their use of the Personal licence. Likewise, most respondents—even those working at medium-sized studios—were either very unclear on, or had not considered, the ramifications of exceeding Unity's $100,000 USD income threshold for videogames published on the Personal licence. They also tended to be unclear on how that income threshold would be enforced or monitored, or whether they had any responsibility to establish a dialogue with Unity Technologies prior to launching their videogames. One team explained that they had 'never had an interaction with a human being from Unity' and that there was no 'vetting process' following the release of their first Unity-developed

---

[9] In many countries, developers of gambling software are required by law to submit their source code to a government regulator. Unity therefore requires that developers of gambling software purchase Unity's underlying source code and use a 'frozen' version of the engine to develop their software. Although the gambling licence generates revenue for Unity Technologies, our contacts were careful to stress that it is not a key growth area for the company.

videogame, which had been developed using a Personal licence. One developer from this team postulated that merely 'imagining' the ramifications of breaking Unity's contractual obligations 'is the panopticon that keeps you in line'. Likewise, some students expressed confusion on Unity Technologies's policy—or lack thereof—for creators who, like them, were earning money from 'pay what you want' donations. It is worth noting, however, that while several respondents were unclear on their contractual obligations, none had been audited. For this reason, most developers spoke in very positive terms about Unity Technologies's 'hands-off' approach to licensing—that is, the way in which the company offers its core toolset for free, yet appears to ask very little in return, thus seeming to uphold its alleged commitment to 'democratizing game development'.[10] When we prompted one developer about how recent changes in Unity's licensing agreement would impact the impending launch of his videogame, he said: 'I don't know. I tried figuring it out and then I just gave up and that seems to be working the best for me. To just not think about it'. Given that few respondents expressed negative sentiments towards Unity, and none had experienced adverse ramifications for not fulfilling their contractual obligations, it seems plausible to suggest that Unity Technologies has deliberately sought to establish a platform ecology wherein users can feel completely undeterred from developing and publishing Unity games, 'to just not think about it'.

To this extent, the bulk of Unity Technologies' revenue comes not from royalties recouped from published content or licensees, but rather from the company's ad network and version control services, as well as venture capital welfare. Much like other platform companies, venture capital has enabled Unity Technologies to rapidly scale up and monopolize its network effects. However, according to informal interviews we conducted with Unity representatives, venture capital is not considered a long-term revenue solution for the company. Instead, Unity Technologies is investing heavily in its ad network services, otherwise known as 'Unity Ads'. An ad network acts as a broker that connects

---

[10] Unity's 'hands-off' approach is not applied universally—one noteworthy case is the Unity-developed mobile videogame *Pokemon Go* (Niantic, 2016), which, when it was first launched, featured a Unity splash screen, suggesting that its developers had developed and published the videogame using the Personal licence. Yet, within days of *Pokemon Go*'s release, the splash screen was removed, suggesting that its developers promptly switched to a different licensing model.

websites and apps that want to host advertisements with large databases of advertisers and advertisements. Ad networks are particularly useful for developers of 'free' mobile videogames. These free mobile videogames often recoup revenue through in-game advertisements and in-app purchases. Developers of these videogames can use Unity Ads to integrate ads natively in their published content. In this arrangement, free mobile videogames—and their developers—are 'brokers' that connect advertisers with eyeballs. Unity Ads provides and enables the underlying ad network, taking a cut from each individual 'transaction' that takes place. Unity Ads is clearly a key element of Unity Technologies's current revenue structure, albeit one that is somewhat beyond the scope of this book and its analysis.

Unity's Asset Store also falls within the remit of the engine's platform ecology. Much like Apple's App Store, Unity's Asset Store allows users to buy, sell, and/or freely obtain various user-developed (sometimes referred to as 'prefabricated') assets and plugins. An example of a plugin is a 'shader' tool, which determines how images are rendered in Unity-developed content. In the past, developers would need to write custom shaders or make do with an engine's default shaders. Through the Asset Store, users can create, share, and obtain any number of user-developed shader plugins, such as Shader Forge and Shader Amplify, many of which are free or sold for a one-off fee. These plugins—and others like them—hook into Unity's editing interface as drag-and-drop tools and custom control panels. When users purchase or obtain an asset or plugin from the Asset Store, they are also granted a licence to use that asset or plugin in published content. Any user can create an asset or plugin and provide it on the Asset Store,[11] though Unity Technologies takes a cut from sales. Asset Store developers must also agree to a condition stating that they cannot request further acknowledgement or compensation for their asset or plugin after setting an initial one-off fee. For this reason, most of our respondents did not feel the need to provide formal

---

[11]Assets and plugins for Unity can also be distributed between users externally from the Asset Store, and are often made available on a developer's own website or alternative hosting platforms such as Git or Itch.io. The Asset Store remains the primary location to distribute and source assets and plugins due to its convenient availability within Unity's editor and the assurance that assets available on the Asset Store are certified to work with supported Unity versions, whereas a file uploaded to a blog might not have been tested on newer versions of Unity.

acknowledgement for the user-developed assets and plugins they were using in their development practices, beyond any initial compensation.[12]

Freelance Asset Store developers perform a labour role quite unique to software culture. Several of our respondents had considered becoming full-time or freelance Asset Store developers, prior to becoming professional videogame developers. There is a strong incentive among Asset Store developers to provide one's plugins or assets for free or for a tokenistic fee. This is driven by something of an open-source mentality and a desire to 'give back' to the community. Many of the more successful user-developed assets and plugins have been acquired by Unity and incorporated into the engine's default toolset. Once a particular asset or plugin becomes popular, as one respondent explained, 'there's a sense of anticipation [...] where everyone's a bit like, "oh, I really want support for this thing, but if I wait like a month, I'll get it properly in the engine"'. As flagged in the introduction, Unity's core affordances are, in this sense, relational; they are shaped as much by the decisions of the company's key stakeholders and in-house software engineers as they are by Asset Store developers and other actors. In the words of one respondent, being an Asset Store developer means continuously providing 'support labour' for one's assets and plugins, in that Asset Store developers 'make their money once, but then they're forced to do a ton of work with every Unity update because Unity keeps breaking things. It's like never-ending work for the plugin maker [...] so eventually they abandon their plugins'. Our respondent compared this situation to a 'Silicon Valley startup model' wherein many Asset Store developers commit to providing unpaid support labour for their assets and plugins in the hope that Unity will eventually buy them out.

One of the effects of 'platform capitalism' (Srnicek 2016) and the 'platformization of cultural production' (Nieborg and Poell 2018) is that users are—knowingly or otherwise—deeply susceptible to software

---

[12]An interesting exception was the director of a Melbourne-based independent studio, who explained that his studio kept records of the various user-developed plugins they had downloaded, in addition to scripts downloaded from programmers' blogs. Whenever his team implements a user-developed plugin or script, they leave a reminder in their code. Once they are approaching launch, they will trawl through the code, contact plugin developers individually, and ask whether those developers require any further acknowledgement or compensation. He explained that, in addition to being 'good practice', this is also a means by which to keep track of any user-developed plugins and scripts that become unstable or, at worst, obsolete as a result of Unity's software updates.

updates, platform policy updates, and interface adjustments. Unity is no exception; its frequent software updates can present complications not only for Asset Store developers but also videogame developers, students, and educators. Unity Technologies frequently updates their core engine software with new features or bug fixes, but these updates can create various instabilities in projects developed on older builds of Unity. To borrow Whitson's (2018a) terminology, many interviewees described Unity updates as moments when the engine exhibited weird, recalcitrant, or voodoo qualities, as if updating may open something of a Pandora's box. As discussed above, given that many respondents were quite reliant on user-developed plugins, they described purposefully not updating Unity unless they had performed significant testing prior to updating. At Melbourne-based videogame studio League of Geeks, studio director Trent Kusters explained that whenever his studio plans to adopt a new Unity version, a software engineer will first quarantine the videogame's current build on a separate machine to see what breaks.[13] Following this quarantine process, the studio then utilizes their in-house expertise as well as Asset Store resources to develop custom solutions and workarounds. However, as Trent explained, this approach can carry the risk of creating a 'Frankenstein' engine that may become unstable in unforeseen ways. Trent's biggest fear—which was shared by several respondents—was encountering a critical issue shortly before the launch of a new videogame. 'Then', he explained, 'if you don't have [the Plus or Pro license], if you don't have source code access, you're just like another pleb in the support forums, being like, "hi, please, we're shipping in two weeks!"'. Students and educators were impacted by Unity updates in similar ways. Frequent Unity updates necessitate frequent curriculum updates, as well as updates to university computers. Despite these risks and complications, most respondents were nevertheless content with using Unity, as the alternatives—developing a custom engine,

---

[13]It is also worth noting that Unity developers cannot take a specific fix from a new update and implement that fix in an older build of Unity, because that would require access to the engine's underlying source code. League of Geeks also explained that, as a studio, they collectively decided not to download plugins from the Asset Store unless those plugins came with source code access. The reason for this is that if a plugin becomes unstable, the studio's software engineer can make the necessary fixes without having to rely upon the plugin creator to provide support.

paying for a licence, or switching to a different engine ecosystem—were, on the whole and for different reasons, not considered viable options.

Through its platform ecology, Unity aims to support and cultivate what Orlando Guevara-Villalobos (2011: 2) calls 'communitarian practices'—practices once associated with grassroots videogame-making communities—that developers associate with openness, care, community, collaboration, and creativity (cf. Parker and Jenson 2017). Many respondents spoke of the various (user-created) online resources available to Unity developers, such as forums, blogs, social media groups, live streams, and video guides. 'The invaluable thing about Unity', as videogame developer and educator Cherie Davidson put it, 'isn't so much the engine itself as how accessible it is to just Google, "how do I do 'x' with Unity", and there will be an answer for you'. Some Unity developers even post their scripts online and invite other developers to use and/or iterate upon them. There are also Unity events—such as the annual 'Unite' conferences, held in Los Angeles, Hyderabad, Berlin, Melbourne, and Singapore—where company representatives, developers, independents, and students come together to network, share knowledge, and discuss the latest Unity features. Cherie interpreted Unite 'as Unity's way of being like, "We are your friends. Talk to us. We're ready to have a dialogue and see what the engine can do for you"'. By actively promoting the formation of what she termed a 'Unity family', Unity's platform ecology is identified not only as a software environment or business model but also a space of affective intermediation where, to borrow Felan Parker and Jennifer Jenson's (2017: 877) words, 'smaller, less business-minded game makers [can] congregate, commiserate, and celebrate creativity'. Unity's ability to establish a relationship of trust with its community—such that the Unity user base not only embraces the open sharing of knowledge but also expects its collective intelligence to be absorbed into the engine's default toolset—is a far cry from the closed proprietary structures and hierarchical business models that once characterized mainstream studio development.

Although Unity is an immensely popular game engine, its long-term commercial viability is not inevitable or guaranteed. By 2020, for example, Unity Technologies reportedly aims to go public via an initial public offering, meaning that it will become more accountable to its investors and shareholders and, perhaps as a result, less driven by its current mission statement to 'democratize game development' (Castillo 2019). Likewise, in the coming years, Unity aims to adopt an entity-component and data-oriented design system in place of its current component-oriented design system. Unity's design system will be explained in more detail in Chapter 3, but the important point here is that Unity's underlying architecture is in a permanent state of flux, and planned changes to this architecture may compromise the perceived 'user-friendliness' of the engine as it currently stands. Regardless of how Unity changes in the future, however, the engine has had, and will have, an undeniably lasting impact on videogame production workflows, design methodologies, and software literacies. The following chapters aim to explore these impacts in more detail, by first zooming in on the specific workflow and grain that Unity brings to bear on videogame development.

REFERENCES

Anthropy, Anna. 2012. *Rise of the Videogame Zinesters: How Freaks, Normals, Amateurs, Artists, Dreamers, Dropouts, Queers, Housewives, and People Like You Are Taking Back an Art Form*. New York: Seven Stories Press.

Banks, John. 2013. *Co-Creating Videogames*. New York: Bloomsbury.

Barton, Matt, and Bill Loguidice. 2009. "The History of the Pinball Construction Set: Launching Millions of Creative Possibilities." Gamasutra, February 6. https://www.gamasutra.com/view/feature/132316/the_history_of_the_pinball_.php.

Bogost, Ian. 2006. *Unit Operations: An Approach to Videogame Criticism*. Cambridge: MIT Press.

Castillo, Michelle. 2019. "Unity Technologies Targeting 2020 IPO: Sources." Cheddar, February 11. https://cheddar.com/media/unity-technologies-targeting-2020-ipo-sources.

Darchen, Sébastien. 2015. "'Clusters' or 'Communities'? Analysing the Spatial Agglomeration of Video Game Companies in Australia." *Urban Geography* 23 (2): 202–222.

Dyer-Witheford, Nick, and Greig de Peuter. 2009. *Games of Empire: Global Capitalism and Video Games*. Minneapolis: University of Minnesota Press.

Freedman, Eric. 2018. "Engineering Queerness in the Game Development Pipeline." *Game Studies* 18 (3). http://gamestudies.org/1803/articles/ericfreedman.

Gazzard, Alison. 2014. "The Intertextual Arcade: Tracing Histories of Arcade Clones in 1980s Britain." *Reconstruction* 14 (1). http://reconstruction.eserver.org/Issues/141/Gazzard.shtml.

Gillespie, Tarleton. 2017. "The Platform Metaphor, Revisited." Culture Digitally, August 24. http://culturedigitally.org/2017/08/platform-metaphor/.

Gitelman, Lisa. 2006. *Always Already New: Media, History, and the Data of Culture*. Cambridge: MIT Press.

Guevara-Villalobos, Orlando. 2011. "Cultures of Independent Game Production: Examining the Relationship Between Community and Labour." In *Proceedings of DiGRA 2011 Conference: Think Design Play*, 1–18.

Haas, John. 2014. "A History of the Unity Game Engine." Interactive Qualifying Project. Worcester Polytechnic Institute. https://web.wpi.edu/Pubs/E-project/Available/E-project-030614-143124/.

Helgason, David. 2014. "Leading Unity into the Future." *Unity Blog*, October 22. https://blogs.unity3d.com/2014/10/22/leading-unity-into-the-future/.

Jenkins, Henry. 2006. *Convergence Culture: When Old and New Media Collide*. New York: New York University Press.

Jørgensen, Kristine. 2017. "Newcomers in a Global Industry: Challenges of a Norwegian Game Company." *Games and Culture* (OnlineFirst). https://doi.org/10.1177/1555412017723265.

Keogh, Brendan. 2019. "From Aggressively Formalised to Intensely In/Formalised: Accounting for a Wider Range of Videogame Development Practices." *Creative Industries Journal* 12 (1): 14–33.

Kerr, Aphra. 2017. *Global Games: Production, Circulation and Policy in the Networked Era*. New York: Routledge.

Kirkpatrick, Graeme. 2013. *Computer Games and the Social Imaginary*. Cambridge: Polity Press.

Kirkpatrick, Graeme. 2015. *The Formation of Gaming Culture: UK Gaming Magazines, 1981–1995*. New York: Palgrave Macmillan.

Lanier, Liz. 2019. "Former Unity Exec Files Lawsuit Alleging CEO Sexually Harassed Her, Others." *Variety*, June 8. https://variety.com/2019/gaming/news/former-unity-exec-files-lawsuit-alleging-ceo-sexually-harassed-her-others-1203236756/.

Lipkin, Nadav. 2013. "Examining Indie's Independence: The Meaning of 'Indie' Games, the Politics of Production, and Mainstream Cooptation." *Loading* 7 (11): 8–24.

Lowood, Henry. 2016a. "War Engines: Wargames as Systems from the Tabletop to the Computer." In *Zones of Control: Perspectives on Wargaming*, edited by Pat Harrigan and Matthew Kirschenbaum, 83–106. Cambridge: MIT Press.

Lowood Henry. 2016b. "Game Engine." In *Debugging Game History: A Critical Lexicon*, edited by Henry Lowood and Raiford Guins, 203–210. Cambridge: MIT Press.

Montfort, Nick, and Bogost, Ian. 2009. *Racing the Beam: The Atari Video Computer System*. Cambridge: MIT Press.

Ng, Benjamin Wai-ming. 2009. Consuming and Localizing Japanese Combat Games in Hong Kong. In *Gaming Cultures and Place in Asia-Pacific*, edited by Larissa Hjorth and Dean Chan, 83–101. New York: Routledge.

Nicoll, Benjamin. 2019. *Minor Platforms in Videogame History*. Amsterdam, the Netherlands: Amsterdam University Press.

Nieborg, David B., and Shenja van der Graaf. 2008. "The Mod Industries? The Industrial Logic of Non-market Game Production." *European Journal of Cultural Studies* 11 (2): 177–195.

Nieborg, David B., and Thomas Poell. 2018. "The Platformization of Cultural Production: Theorizing the Contingent Cultural Commodity." *New Media & Society* 20 (11): 4275–4292.

Nooney, Laine. 2013. "A Pedestal, a Table, a Love-Letter: Archaeologies of Gender in Videogame History." *Game Studies* 13 (2): n.p. http://gamestudies.org/1302/articles/nooney.

O'Donnell, Casey. 2014. *Developer's Dilemma*. Cambridge: MIT Press.

Parikka, Jussi. 2012. *What Is Media Archaeology?* Cambridge: Polity Press.

Parker, Felan, and Jennifer Jenson. 2017. "Canadian Games Between the Global and the Local." *Canadian Journal of Communication* 42: 867–891.

Sack, Warren. 2019. *The Software Arts*. Cambridge: MIT Press.

Sotamaa, Olli. 2010. "Play, Create, Share? Console Gaming, Player Production, and Agency." *Fibreculture Journal* 16. http://sixteen.fibreculturejournal.org/play-create-share-console-gaming-player-production-and-agency/.

Srnicek, Nick. 2016. *Platform Capitalism*. Malden: Polity Press.

Stuckey, Helen, Melanie Swalwell, Denise de Vries, and Nick Richardson. 2015. "What Retrogamers Can Teach the Museum." In *MWA2015: Museums and the Web in Asia*. https://mwa2015.museumsandtheweb.com/paper/what-retrogamers-can-teach-the-museum/.

Švelch, Jaroslav. 2018. *Gaming the Iron Curtain: How Teenagers and Amateurs in Communist Czechoslovakia Claimed the Medium of Computer Games*. Cambridge: MIT Press.

Swalwell, Melanie. 2012. "Questions About the Usefulness of Microcomputers." *Media International Australia* 143 (1): 63–77.

Takahashi, Dean. 2018. "John Riccitiello Q&A: How Unity CEO Views Epic's Fortnite Success." Venturebeat, September 15.

Tkacz, Nathaniel. 2014. *Wikipedia and the Politics of Openness*. Chicago: The University of Chicago Press.

Turkle, Sherry. 2005 [1984]. *The Second Self: Computers and the Human Spirit*, 20th Anniversary Edition. Cambridge, MA: MIT Press.

van der Graaf, Shenja. 2018. *ComMODify: User Creativity at the Intersection of Commerce and Community*. New York: Palgrave Macmillan.

Vogel, Michael. 2017. "Japanese Independent Game Development." MA dissertation, Georgia Institute of Technology. https://smartech.gatech.edu/bitstream/handle/1853/58640/VOGEL-THESIS-2017.pdf?sequence=1&isAllowed=y.

Whitson, Jennifer R. 2018a. "Voodoo Software and Boundary Objects in Game Development: How Developers Collaborate and Conflict with Game Engines and Art Tools." *New Media & Society* 20 (7): 2315–2332.

Whitson, Jennifer R. 2018b. "What Can We Learn From Studio Studies Ethnographies? A 'Messy' Account of Game Development, Materiality, Learning, and Expertise." *Games and Culture* (OnlineFirst). https://doi.org/10.1177/1555412018783320.

CHAPTER 3

Workflow: Unity's Coordination of Individualized Labour Processes

Abstract This chapter considers how cultural software situate and mediate the labour processes and skills of their users. Unity, the chapter argues, positions itself as a *metaplatform* for the coordination of workflows that are intensely individualized and distributed across multiple software environments. The chapter begins with a close analysis of Unity's software environment and a consideration of how it encourages a 'component-oriented' approach to design. It then considers how this approach redirects the videogame development pipeline, decentring the role of the programmer in the videogame development team. Rather than simply 'empowering' non-programmers, however, the final section considers how Unity's coordination of workflows requires developers to streamline, coordinate, and individualize their own labour processes, thus contributing to a broader reimaging of creative work under capitalism.

Keywords Workflow · Labour · Unity game engine · Platformization of cultural production · Skills · Videogame development

This chapter considers how cultural software situate and mediate the labour processes, skills, and workflows of their users. Workflow refers to the ways that developers streamline, coordinate, and also individualize their labour processes with and through software. Workflow is a term with strong affinities to software culture, albeit one whose genealogy can be traced back through corporate discourses, domestic science, and self-help guides focused on productivity and time management (see Gregg 2018). Workflow is consonant with broader trends of self-sovereignty in software culture, in that an individual's workflow—what software they use and how they combine methods and techniques across software environments—amounts to a unique portfolio or personal style. When any given software project requires the coordination of multiple work-flows—those of programmers, designers, and artists, for example—they coalesce into a development pipeline. As discussed in the introduction to this book, the traditional role of a game engine is to coordinate a team's workflows and make them interoperable, such that they can be fed into a pipeline (Banks 2013; O'Donnell 2014; Whitson 2018). Unity in particular positions itself as a metaplatform for the coordination of work-flows that are intensely individualized and distributed across multiple software environments. As one respondent put it, 'just as important as learning Unity is learning the workflow of how you should be using it— or *a* workflow, I should say, because everyone is probably different'.

Conceivably, there exist just as many Unity workflows as there are Unity developers. The workflow of a level designer in Unity will differ from that of an animator, just as the workflow of a programmer will differ from that of an audio engineer, and so on. Workflow can be shaped by disciplinary expertise, the suite of software tools used by an individual or team, or even the placement of control panels and windows in a software interface. Unity's performed platform neutrality and its ability to act as a site of 'deep remixability' (Manovich 2013: 268) for differentiated software techniques makes pinpointing its direct impact on any one specific workflow difficult. The question, then, is not whether Unity imposes a particular workflow on developers, but rather how it enables different relationships between multiple possible workflows. Central to such a question is how Unity distributes power dynamics among a videogame development team, how it sets a template for the requisite skills and roles necessary for videogame development, and how it leverages its status as a metaplatform to enrol a diversity of workflows into its software ecology.

Videogame development, as a set of skills, has long been poised uncomfortably between computer programming and creative practice. Historically, programmers and software engineers have been central to the process of videogame development—first as hackers and tinkerers who produced early computer games, and later as professionals in a formalized industry where they were (and often still are) responsible for constructing and maintaining the code frameworks that undergird studio pipelines. While the shift from programmer-centric development to content-centric development in the 1990s implied that the workflows of videogame designers, artists, writers, and sound engineers were just as (if not more) important as those of programmers and software engineers, most in-house production workflows, tools, and pipelines remained implicitly catered towards the disciplinary expertise of the latter disciplines. For this reason, programmers and software engineers often maintained some level of formal or informal authority over pipelines, acting as gatekeepers and bottlenecks through which content had to be approved and implemented. Key, then, to Unity's supposed democratization of videogame development is its ability to decentre the authority of the programmer and, ostensibly, to make it possible for designers and artists to participate more seamlessly in the development process.

This chapter starts with a thick description of Unity's project structure and editor interface to demonstrate how Unity encourages component-oriented, as opposed to object-oriented, workflows. The second section considers the consequences of Unity's decentring of the programmer. For a small development team or studio, the ability to use Unity (and its robust network of support and resources) in place of a dedicated engine programmer or team of engine programmers may seem liberating from a financial and creative standpoint. Yet, by imposing its cultural (software) framework on existing team compositions, Unity is complicit in a broader individualization of creative labour in software culture, wherein media creatives are expected to 'find their way' in an environment devoid of job security, welfare support, and collective organization. The chapter thus concludes with a discussion of how Unity accommodates digital cultural work within a broader 'creativity *dispositif*' (McRobbie 2016) that articulates cultural work as commodified and individualized.

UNITY'S COMPONENT-ORIENTED DESIGN SYSTEM

A single Unity project works at a series of scales: project, scene, game object, and component. Project refers to the entire project being developed. A Unity project houses all the files associated with that project—2D sprites, 3D models, scripts, texture materials, audio files, and so on—each of which are commonly made using other cultural software and imported into the Unity project. A single project is composed of a number of scenes. Following film terminology, a scene is a single unit of the project, such as a particular level of a videogame, or a test scene wherein a developer can try out new mechanics. Smaller videogames might exist entirely in a single scene, but larger videogames will have many more. A single scene also contains a number of game objects. A game object could be anything from the player avatar, the entire simulated environment, the virtual camera, a weather system, screen elements such as a health metre or pause menu, or an invisible object keeping track of different variables.

Every game object is itself a bundle of components, each of which determine some aspect of that game object. For instance, every game object has a 'transform' component, which determines the position, rotation, and scale of the object within the scene's Cartesian coordinate space. A 2D sprite game object would have a 'sprite renderer' component, which determines aspects of how the 2D sprite is rendered, whereas a 3D object would have a mesh renderer, which renders it as a three-dimensional object. Collider components determine if the game object is solid, while rigidbody components determine if and how the object is impacted by simulated gravity. Users are able to make new components via scripts or download existing ones from the Unity Asset Store. For instance, a simple 'move' script might allow an object to move in a straight line towards the player's character. This script could be attached to any number of objects as a component. It could be edited to move at a slow speed on one object, but then at a faster speed on another. This is the general hierarchy of Unity development: a project is split into scenes; a scene consists of a number of game objects; a game object consists of a number of components.

As illustrated in Fig. 3.1, the screen space of the Unity editor is separated into a number of panes, each of which can be repositioned and resized as per a particular user's workflow. While there are many possible panes to have visible at any given time, five panes are the most common and core for a basic comprehension of how Unity functions: project, hierarchy, scene, game, and inspector. The project pane functions as an explorer-type folder structure of all the files available in the project. The hierarchy pane lists all the game objects that exist in the currently open scene. The inspector pane provides a list of all the components attached to the currently selected game object. Finally, the scene and game panes provide two different perspectives on the currently active scene.

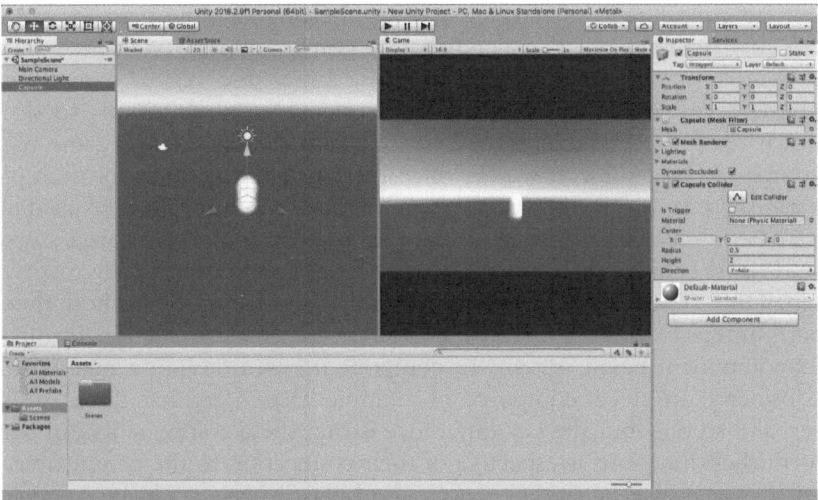

Fig. 3.1 A new scene in the Unity editing interface (version 2018.2.9f1). Unity and Unity logos are registered trademarks and trademarks of Unity Technologies or its affiliates in the USA and elsewhere. All rights reserved. Other images, content, names, or brands are proprietary of their respective owners. Neither this book nor its authors are affiliated with, or endorsed or sponsored by, Unity Technologies or its affiliates

The scene pane provides a workspace view of the scene where objects can be placed, edited, deleted, and moved, but with effects such as the sky-box and lighting disabled. The game pane, on the other hand, provides a view of the game space but from the view of what the hypothetical player would see. The game pane is most crucial when considered in relation to Unity's 'play' button. By pressing the play button, the user can immediately, without leaving the Unity editor, play the scene in its current state in the editor while continuing to tweak objects and variables on the fly.

While Unity's core infrastructure relies on object-oriented programming languages such as C#, its editor promotes what some developers call a component-oriented design system. We discussed this term with Anthony Coculuzzi, a Canadian programmer who was working temporarily at The Arcade when we conducted our interviews. Anthony distinguished component-based design from object-oriented programming using the metaphor of a bike. In object-oriented programming, a bike can be considered an object constructed from multiple parts or components, such as a seat, a handlebar, a gearset, and so on. These parts contain properties that extend the base states and related behaviours of the bike; brake pads, for example, cause the bike's wheels to slow down once the brake levers are depressed. Likewise, a videogame may contain a player object, which, once broken down, consists of multiple parts such as an inventory, health, model, and so on. A power-up item may be implemented in such a way that it alters the player object's base states and behaviours. Object-oriented programming essentially involves separating objects into their most basic parts and considering how those parts interact. Unity promotes a different approach to design through its component-based system. In component-based design, a bike is still considered an object composed of multiple parts—seat, handlebar, gearset, and so on—but the key difference is that these components can perform their functions irrespective of their connection to the parent object (that is, the bike). In other words, the bike's wheels could be removed and attached to a wheelchair, and they would still perform the same basic function.

Unity's component-oriented design system—and the ability to augment that system with custom workflow solutions and plugin capabilities—enables developers to create individualized toolkits that, much like proprietary engines, can be used to iterate on existing content and develop a unique toolchain. Consider, for example, the Unity-developed project *Virtual Songlines* (First Nations Software, 2018), led by developer Brett Leavy. Leavy and his team have used Unity to create a toolkit of assets and components that can be used on multiple projects for the same underlying purpose: to recreate Australian landscapes as they were prior to British colonization, from the perspective of First Australians. The *Virtual Songlines* toolkit includes hundreds of assets based on Australian flora and fauna, as well as a design structure that derives knowledge from the cultural heritage of Aboriginal landowners. It allows different environments to be constructed relatively quickly from this library of pre-existing elements. Brett explained that he can present this toolkit to potential clients interested in having their community virtually recreated, much as he would a proprietary tool. Here, the 'Unity-ness' of the tool fades into the background, and *Virtual Songlines* is instead conceptualized as a quasi-proprietary toolkit in itself.

While object-oriented programming necessitates a detailed understanding of an object's functionalities, component-oriented design can be less 'clean' by comparison. For Anthony, 'you can get really lost in how things are linked together or how things are supposed to work together' in Unity's component-oriented design system. For instance, one could alter the aforementioned 'wheel' component to make it work better on bike objects, while inadvertently overlooking its connection to (and implications for) wheelchair objects. Component-oriented design is nonetheless advantageous for Unity developers who are not proficient programmers. By enabling developers to drag and drop different objects, components, and files between Unity's different panes, tweak the settings of an object's components in the inspector pane without having to edit code directly, and test prototypes in the editor without first requesting a build from the programming team, Unity arguably levels out the hierarchies inherent in traditional videogame development pipelines and affords more individualized workflows.

DECENTRING PROGRAMMERS, REDIRECTING WORKFLOWS

One of the key claims made by and about Unity is that it empowers designers and artists to make contributions to projects without having to default to the technical expertise of a programmer or software engineer. Consider, for example, the Unity-developed videogame *The Gardens Between* (The Voxel Agents, 2018), published in 2018 by a small team of developers known as The Voxel Agents. *The Gardens Between* is a side-scrolling puzzle videogame with a distinctive visual aesthetic and time-manipulation mechanic. Matthew Clark, programmer and co-founder of The Voxel Agents, explained that he collaborated with a broad team of designers, writers, and artists on the project, many of whom had never worked on a videogame before. Through Unity, each member of the team was able to directly create and edit content, without needing to filter their decisions through a programmer's skill set, taste, and workflow. By the end of the project, the animator, who had never worked on videogames before, had taught himself shader programming and was even writing custom scripts using a visual scripting plugin called Playmaker. 'We couldn't have made *The Gardens Between* without his ability to do that', Matthew told us, 'because I'm not an artist. I'm a programmer'. The Voxel Agents are one among many teams who have leveraged tools such as Unity to adopt a decentralized team structure, wherein programmers, designers, and artists are afforded the ability to individualize and thus authenticate their workflows. 'Back in the day', Matthew explained, '[shader programming] was just a really scary thing to even attempt to learn. There was just so much technical stuff that you would need to wrap your head around in order to understand shader programming, whereas, just the fact that the [Unity] tools make it easy to get that feedback in real time makes it easier to teach yourself'. Having foregone the hierarchical production pipelines that characterized videogame development in the 2000s and early 2010s, teams such as The Voxel Agents are using Unity to make it easier for artists, designers, and audio engineers to directly participate in the collaborative process.

Reflecting on the advantages of a decentralized team structure, Matthew recounted his experiences working in the Australian videogame industry prior to the widespread adoption of Unity. He provided an example of a proprietary tool that had artists and designers implementing changes to a videogame's user interface by directly editing a text file. 'If you wanted a button [on the user interface] to be a bit further to the left', he explained, 'you'd have to [type out] x = −12, and then our system would compile that into the build and then make the build and then you'd play the build [to check it], so you couldn't do any live editing'. To this extent, one of Unity's key impacts on production workflows—and design methodologies by extension—is that it offers real-time feedback on edits through its component-based design system and inspector pane. In most proprietary engines, developers need to implement changes (usually via the expertise and at the whim of a programmer), compile, test, take screenshots, record necessary changes, and then go back to make the necessary edits. Matthew explained that, when working with custom tools in this way, there was 'always a trade-off between effort and quality—if the effort is too high you'll just stop at a certain quality level because it's not worth it to go higher'. As a programmer, Matthew would often need to implement changes and requests made by designers and artists. He described having to make judgements as to the quality of the proposed change versus the effort required to implement said change. 'When designers would request features', he explained, 'if I thought an idea was bad and it was going to be hard work, it's really easy to push back and be like, "no, it's going to be too hard for me to build this tool"'.

In this hierarchical arrangement, which is also documented in Casey O'Donnell's (2014) account of development pipelines in larger studios, the programmer possesses extensive power to determine the direction the videogame takes. When using Unity, however, the ability for different team members to, with minimal effort, place objects in a scene and press play decentres the programmer from the pipeline. If a videogame designer or artist wishes to experiment with different variable values, they can simply press play and try the scene while adjusting the values on different components in the inspector. Furthermore, given that Unity is interoperable with a range of different software tools familiar to designers, artists, and audio engineers—Maya, Blender, FMOD, and so on—it is in a position to enrol a variety of possible workflows. In component-oriented development, there is often less of a singular pipeline constructed from subsequent, individual workflows, and more of a shared pool, with each team member's workflow directly interfacing with the project itself. In this way, individual developers are able to make direct changes and exert a direct agency over the shape and direction of the project.

The Melbourne-based videogame studio House House provide another example of this decentring of the programmer in the development pipeline. House House started their first project, *Push Me Pull You* (2016), as a hobbyist project. Nico Disseldorp, the only member of the team with programming experience, developed the videogame in a custom-made web browser engine. In his own words, Nico acted as a 'bottleneck' through which content was implemented in the game. Once the team switched to Unity for their subsequent project, *Untitled Goose Game* (2019), they each learned distinct parts of the engine—as another member of the team explained, 'someone had to learn the animator component of Unity, so that person became the animator'. This meant that despite the team 'not really having job titles', each member began self-identifying as having a distinct role. Each member was therefore able to directly interface with the project through the Unity editor, rather than wait for Nico's availability to import their changes.

By the same token, House House were one among several teams we interviewed that described encountering more conflicts in the collaborative process as a result of adopting a more decentralized and individualized workflow structure. When multiple team members work directly on the same Unity project, they run the risk of generating merge conflicts. Merge conflicts occur when two or more simultaneous edits on the same project file are incompatible or contradictory. It is then necessary to pick through past iterations of code by using version control software such as Git to identify and resolve said conflicts. Most teams we spoke to were using in-house management processes and custom strategies to alleviate these risks. House House, for example, described using a physical totem that could only be on one member's desk at any given time. 'Unless the totem was on your desk', Nico explained, 'you knew you weren't meant to touch the scene file'. Another developer spoke of structuring his team's code base such that scenes were not interdependent on each other—a smaller scene could be edited without having to edit a broader scene, for example—which enabled his team 'to keep our workflows apart so that we can both work independently without stepping on each other's toes'. Larger studios relied more heavily on the guidance of creative leads and project managers to strategically manage the workflows of individual team members. In this way, Unity decentres the role of the programmer but recentres the role of the creative lead, who is now tasked with coordinating the workflows of their team.

In many cases, Unity does not simply decentre the development team's programmer, but replaces them entirely. Or, put more positively, Unity enables videogame development for teams that lack the resources to hire more specialist programmers. Morgan Jaffit of Brisbane-based studio Defiant Development, for example, explained that if his studio were to build their own engine, they would need to hire multiple software engineers, which would far outweigh the cost of a Unity licence:

> I mean, what is a programmer worth? Not all programmers' salaries, especially in Australia, are huge. But even conservatively, taking a programmer on a $60,000 [AUD] salary, a low salary for most professional programmers. That $2,000 [a year] Unity license accounts for $5,000, $10,000 a month [to hire a programmer]. It's about two weeks of their work? The concept that you can't get more done by having Unity as your starting point than a programmer could achieve for two weeks of work, with the exception of some very, very, very specific game types, is boggling.

In this case, for Morgan, Unity does not simply displace certain programming skill sets but is capable of potentially *re*placing them in some roles. Furthermore, utilizing Unity as a low-cost resource means taking advantage of its general-purpose toolset and its extensive network of support, which includes its Asset Store, online troubleshooting provided by users, and regular software updates and fixes provided by Unity's in-house software engineers. This allows a studio to, in Morgan's words, be a 'games company, not a tech company'. While in the past creating the tools and the videogame were often synonymous, with engines like

Unity, these requirements become more differentiated. In the process, the dynamics and relationships between different roles and workflows within the pipeline are reconfigured.

PRODUCTIVE WORKFLOWS

While many non-programmers speak of being more empowered by the ability to manipulate components in Unity without going through the potential bottleneck of a programmer, this does not mean that Unity is a completely open or unbiased tool. Rather, Unity takes on a mediating role—similar to that traditionally held by the programmer—in the various practices, decisions, and workflows it encourages and discourages. As one developer put it,

> Unity affords specific practices and pipelines and processes, and you can fight against that, but to be most effective and to get the most out of Unity, you need to find ways of working within those constraints and turning the weaknesses into strengths. You've got to find ways of being like, 'oh well, I can't do this in Unity, but I'm going to make a game that doesn't need that'.

This balancing act of working within and against Unity's affordances can be conceptualized through the metaphor of 'grain', which we explore in more detail in the following chapter. In the meantime, it is important to consider how Unity specifically and cultural software generally render opaque their own mediation of their users' workflows. When the above developer described 'making a game that doesn't need' what he could not do in Unity, he was describing a pipeline bottleneck not dissimilar to that once determined by a programmer. Indeed, 'Unity' here comes to stand in for a wider network of programmers—those working for Unity as well as the broader community of Asset Store developers, bloggers, and forum contributors. Unity, then, does not replace the programmer at all, but displaces them to a background, external position, beyond the usual conceptualization of the development team.

Thus, rather than directing their frustrations at a programmer, developers direct their frustrations at Unity's restrictions, assumptions, and preferences—or what Whitson (2018) calls the engine's voodoo agency. Game engines, as Whitson (2018) argues, are often viewed as recalcitrant, opinionated, and unpredictable, never fully bending to the expectations of their users. For Whitson, this often means that game engines reinscribe the authority of the programmer, as designers and artists often find themselves having to summon the programmer's magic touch to 'coax' the engine 'into alignment' with their creative visions (Whitson 2018: 2316). Likewise, many developers speak of coaxing Unity into alignment with *their* creative workflows as opposed to working with or against *Unity's* specific workflows. Ironically, many Unity developers conceptualize the labour of videogame development through a programming lens, where the software is not considered a mediating tool that makes its presence felt, but rather as a neutral platform that fades into the background as a means of enabling seemingly autonomous creativity.

This connects to a particularly neoliberal and post-industrial fetishization of 'creativity' that 'now appears to value more "flexible", "aesthetic", and "soft" workplace cultures' (Banks 2007: 92) in ways that are particularly exemplified by the tech industry (Gregg 2018). This fetishization, in turn, individualizes cultural work as a means of disassociating creative labour 'from traditional notions of what might make a good workplace (order, planning, efficiency, democracy, mutuality, security and stability)' (Banks 2007: 92; see also McRobbie 2016). Unity's enrolment of a wider range of workflows can thus be contextualized within the critical literature on cultural work, which demonstrates how creative labour is increasingly tied to notions of self-actualization and 'meaningful' work in place of traditional social ties and collective forms of support (Banks 2007; McRobbie 2016; Gregg 2018). In cultural software, the practice of the commodified cultural worker and the workflow of the entrepreneurial tech worker converge in a desire for self-governed productivity.

Unity's aim to democratize development may indeed empower people with different skills to make and share videogames. However, those people are typically doing so with less resources than those that can afford to make or lease proprietary engines, and without the security and benefits of traditional employment structures. Unity's conception of 'democratized game development' contributes to a romanticized and individualized notion of creative work under neoliberalism that commonly sugar coats precarious, contingent work, or a lack of employment opportunities under the utopic language of 'entrepreneurism' (Oakley 2014). In Australia, for instance, after the obliteration of an industry dependent on the resources of publishers based in North America, Unity facilitated the emergence of small, independent teams who had no alternative pathways to economically sustainable videogame development. In facilitating such teams, Unity fits into an increasingly entrepreneurialized environment of videogame development, where, as Guevara-Villalobos (2014: 733) explains, 'independent developers need to administer their budgets, self-regulate their working hours, keep themselves updated with knowledge and information about business models, general industry, new applications and technologies, as well as take charge of marketing and public relations'. Much of this 'relational labour' (Baym 2015) involves 'interfacing' (Whitson et al. 2018) with the wider indie ecosystem—that is engaging festival attendees, Twitter followers, videogame critics, powerful cultural intermediaries, and institutional gatekeepers—in an effort to improve the marketability and discoverability of one's project. Many Unity developers are what Lobato and Thomas (2015: 49) call 'necessity entrepreneurs' as opposed to 'opportunity entrepreneurs'. Where opportunity entrepreneurs are 'those who can see and act on market opportunities, and are the classic self-starting go-getters', necessity entrepreneurs 'are locked out of certain markets [and] improvise and get by however they can. Everyday work takes on a quality of individual, ad hoc enterprise, whether they like it or not'. This distinction is crucial and commonly not made in discourses around both cultural and technological work. People performing precarious and contingent work (such as Australian videogame developers using Unity), doing what they can do to 'get by', are instead romanticized as self-driven and autonomous opportunity chasers.

As flagged in the introduction, Unity is characterized by a contradiction of empowerment and disempowerment in the workflows it affords. It is a useful illustration of Wendy Chun's (2011: 59) argument that software interfaces 'have become functional analogs to ideology and its critique'. Broadly speaking, ideology—in its various incarnations—aims to naturalize worldviews and make opaque hegemonic structures appear commonsensical. Likewise, software interfaces make complex computer processes (which ultimately conceal structures of power) appear 'transparent', and they provide tools for authoring and editing as a means of bolstering 'our seemingly sovereign—empowered—subjectivity' (Chun 2011: 9). Unity simplifies and demystifies, but in the process, it invariably re-enforces a particular vision of how videogame content should be developed. 'You can fight against [Unity]', as the above respondent put it, 'but to be most effective and to get the most out of Unity, you need to find ways of working within those constraints and turning the weaknesses into strengths'. Importantly, however, Chun's (2011: 59) argument is that software interfaces mimic ideology *and its critique*. Tools such as Unity enable a whole range of media creatives to participate in—and thus critique—a craft that was once considered prohibitively complex for non-programmers. To appropriate Chun's (2011: 59) words, Unity is 'a powerful response to, and not simply an enabler of, postmodern/neoliberal confusion'. As she elaborates, 'using free software does not mean escaping from power, but rather engaging it differently' (Chun 2011: 21). This statement is evident in the paradoxical combination of empowerment and disempowerment at the heart of Unity's interface, as well as in the attitudes of the constituents it enrols. Unity does not necessarily give developers tools to escape the videogame industry's legacy power structures and design traditions, but it does at least provide a means by which developers from different backgrounds might engage those structures and traditions differently.

While Unity (and commercial game engines more generally) is arguably a net positive for developers in the way it displaces the previously ingrained dominance of the programmer within the pipeline hierarchy, it also points towards how cultural software generally contribute to a shift in perceptions of cultural work from irrational creativity—art for art's sake—to commodified, rationalized, and individualized 'workflows'. Cultural software empowers—within particular entrepreneurial framings—but it also provides particular shape and structure to the creative work it empowers. While this chapter considered Unity's shaping of developers' production workflows, the next chapter considers its impact on the creative process itself and the cultural reception of Unity-developed videogames.

REFERENCES

Banks, John. 2013. *Co-creating Videogames*. New York: Bloomsbury.
Banks, Mark. 2007. *The Politics of Cultural Work*. New York: Palgrave Macmillan.
Baym, Nancy K. 2015. "Connect with Your Audience! The Relational Labor of Connection." *The Communication Review* 18 (1): 14–22.
Chun, Wendy Hui Kyong. 2011. *Programmed Visions: Software and Memory*. Cambridge: MIT Press.
Gregg, Melissa. 2018. *Counterproductive: Time Management in the Knowledge Economy*. London: Durham University Press.

Guevara-Villalobos, Orlando. 2014. "Artisanal Local Networks: Gamework and Culture in Independent Game Production." In *Handbook of Digital Games*, edited by Marios C. Angelides and Harry Agius, 730–750. Hoboken: Wiley.

Lobato, Ramon, and Julian Thomas. 2015. *The Informal Media Economy*. Hoboken, NJ: Wiley.

Manovich, Lev. 2013. *Software Takes Command*. Cambridge: MIT Press.

McRobbie, Angela. 2016. *Be Creative: Making a Living in the New Culture Industries*. Cambridge: Polity Press.

Oakley, Kate. 2014. "Good Work? Rethinking Cultural Entrepreneurship." In *Handbook of Management and Creativity*, edited by Chris Bilton and Stephen Cummings, 145–159. Cheltenham: Edward Elgar.

O'Donnell, Casey. 2014. *Developer's Dilemma*. Cambridge: MIT Press.

Whitson, Jennifer R. 2018. "Voodoo Software and Boundary Objects in Game Development: How Developers Collaborate and Conflict with Game Engines and Art Tools." *New Media & Society* 20 (7): 2315–2332.

Whitson, Jennifer R., Bart Simon, and Felan Parker. 2018. "The Missing Producer: Rethinking Indie Cultural Production in Terms of Entrepreneurship, Relational Labour, and Sustainability." *European Journal of Cultural Studies* (OnlineFirst): https://doi.org/10.1177/1367549418810082.

CHAPTER 4

Grain: Default Settings, Design Principles, and the Aura of Videogame Production

Abstract This chapter considers Unity's aesthetic impact on the cultural work it is used to produce, developing the metaphor of 'grain', from woodworking, to consider how users are oriented towards particular design methodologies. The chapter first considers how videogames inherit a nebulous 'look and feel' from their engines. It then charts the interrelationship between the design principles that are developed within a medium and the design standards that are imposed by software and that, in Unity's case, *become* design principles. Finally, the chapter considers Unity's dominant design standard of iterative design, which encourages Unity users to take advantage of existing assets. Here, the look and feel of a game engine is directly caught up with the aura, or lack thereof, of an individual videogame work.

Keywords Unity game engine · Look and feel · Game feel · Grain · Craft · Videogame development

Like all cultural software, Unity has default settings and processes that attempt to pre-empt the most common workflows through which users—in this case, videogame developers—create, edit, and iterate upon content. While Unity does offer those with the requisite skills the opportunity to make low-level adjustments and manipulations in the engine's editor environment, many developers use Unity explicitly because they lack such a skill set. Thus, most Unity developers feel compelled to adopt design methodologies that approximate Unity's 'opinion' of what the design process should look like—to, in the words of several respondents, 'follow the path of least resistance'. Many respondents also emphasized the need to actively and deliberately divert from this path in their workflow and design, lest their final product be too readily identifiable as just another 'Unity game'.

In this chapter, we consider Unity's aesthetic impact on the cultural work produced with it using the metaphor of 'grain'. Several respondents used the term grain when reaching for a way to describe Unity's impact on their work, and we find it a fruitful concept for considering the relationality of a cultural software's affordances (see Bucher and Helmond 2017). Going with or against the grain of something is a colloquialism derived from woodworking—that is, cutting with or against the pattern and orientation of wood fibres. In a very abstract sense, Unity also has 'patterns' and 'fibers'—protocols, standards, and affordances—that orient users towards particular design methodologies. Grain thus allows us to consider how design processes enter into dialogue with broader cultural processes. For game engines in particular, grain articulates how videogames inherit a particular 'look and feel' from their engines that is then received either positively or negatively by an audience. As a concept, grain also draws attention to the banal but often overlooked materiality of digital creative work. One studio director put it thus:

> Without getting all Marshall McLuhan about it, the medium matters and the artistic medium of creation matters, and what the sorts of toolsets enable you to do easily matter. It even comes down to the sorts of visual things that are present. The filters and focuses and ways of putting polygons on screen that are easily there and easy to use, the fonts and the text; there's a definite look to a Unity game. Or there was. Not all games look like Unity games, but there's a definite kind of baseline look.

Like any creative tool—be that a canvas, camera, or musical instrument—cultural software have ways in which they are easier to use and ways in which they are more difficult to use. A particular cultural software will make certain design decisions more likely and others less likely. However, unlike analogue tools such as musical instruments, cultural software are also bound by software protocols that *enforce* certain design decisions—they are, in the words of one respondent, 'one of the of the only creative methods that tells you "no"'. This is not to say that cultural software 'determine' the creative decisions of their users in any straightforward or linear manner, but rather that cultural software have distinct grains that orient developers towards particular design standards and methodologies (Norman 2013). As flagged in previous chapters, it is also important to note that Unity's grain is relational, in that it is shaped not only by the decisions of Unity Technologies's key stakeholders and in-house software engineers, but also by the collective intelligence of its online communities and Asset Store developers, whose assets and plugins are regularly folded into Unity's default toolset.

While Unity makes claims to empowerment and democratization by virtue of being a general-purpose game engine, most respondents regarded Unity's grain as annoying, presumptuous, or limiting. This speaks to the (long-standing) fantasy of a blank, neutral game engine through which 'any' videogame idea can be realized, and upon which 'any' workflow can be utilized. Ironically, while respondents regarded Unity's grain as stifling, they tended to view more specialist, niche, or grassroots game engines in a more positive light. The highly defined affordances of, for example, the Twine engine, which can only be used to create interactive fiction, or Bitsy, which restricts developers to 8×8 pixel sprite images and a three-colour palette, are viewed as quintessential to what a 'Twine game' or a 'Bitsy game' fundamentally *is* and ultimately as conducive to creative expression. Unity's grain, meanwhile, is a locus of popular anxieties around issues such as 'asset flipping' (a derogatory expression describing videogames built from prefabricated assets) and the supposed 'indiepocalypse' (the fear that, as a result of the various barriers to videogame development being lowered, there are now 'too many games and too many developers' [Wright 2018: n.p.]).

These anxieties essentially boil down to a fear that Unity's component-oriented design system (explored in the previous chapter) renders it 'too easy' for people to make videogames. The notion that there are too many 'Unity games' connotes a glut of interchangeable, mass-produced, and depersonalized videogames constructed from prefabricated elements, as opposed to hypothetically bespoke, 'handcrafted' videogames made from the ground up. There is, among our respondents and in videogame culture more broadly, a general sense that the grain of Unity is too easily recognizable and detectable across videogames produced in Unity. Such claims are highly political, as will be discussed further in the following chapter, but they do point to the fact that Unity has a perceptible 'look and feel' that deeply mediates the cultural production undertaken by its users. Yet, unlike grassroots game engines or, indeed, proprietary engines, it is rare for a Unity videogame to flaunt the fact that it was 'made in Unity'. In order to market itself as a universal platform on which 'any' videogame can feasibly be created, Unity Technologies has a vested interest in ensuring that the grain of its engine remains ambiguous, because by doing so, it naturalizes Unity as the 'default' tool for videogame production.

The first section of this chapter considers what it actually means to claim that a particular game engine has a particular 'look and feel', so as to better contextualize the role of the engine on the play experience. We then consider the relationship between cultural software and developer, by highlighting our respondents' articulations between the 'design principles' of their craft and the 'design standards' that a piece of software such as Unity imposes through its ubiquity. The final section looks at how the grain of Unity influences a standard of iterative design both within a single project but also across Unity projects, as assets, code, and ideas are distributed, reused, and modified between different developers. Here, the processes of designers and the anxieties of players intersect through long-standing debates around handcrafted versus mass-produced art.

THE LOOK AND FEEL OF A GAME ENGINE

In the block quote above, our respondent notes that there is often a discernible look to videogames made in the same game engine, inherent in seemingly mundane elements such as 'ways of putting polygons on screen'. It is common for developers, critics, and players to attempt to identify the engine provenance of a particular videogame through judgements based on intuition and connoisseurship—judgements that are, in most cases, difficult to break into a set of well-defined schemas. The 'look' of a particular game engine can, in some instances, be identified through the visibility or arrangement of particular elements that the player recognizes from other videogames made using the same game engine: the use of, for example, particular fonts, particular methods of rendering lighting or simulating physical interactions, or particular stylings of user interface elements. These elements not only impact how a videogame made in a particular engine looks, but also how it feels. By 'feel', we are referring to videogame play as an embodied experience of watching, listening, and touching, as has been increasingly explored by designers through notions of game feel (Swink 2009) or juice (Gabler et al. 2005; Brown 2016), and by scholars through phenomenological (Keogh 2018) and affect theory (Anable 2018) approaches. A particular videogame feels a particular way to play through the intermingling of mechanics and narrative with controller response times, input devices, muscle memory, animation speeds, play environment, screen size, and a whole range of technical, artistic, and embodied elements.

How a videogame looks, sounds, and controls directly influences how that videogame feels to play, and all three of these aspects are in part influenced by the game engine used. To claim a game engine has a 'look *and* feel' is to draw an explicit, irreducible connection between a videogame's visual and haptic components.[1] A game engine has default ways of simulating physics, responding to inputs, casting light, drawing textures, or generating audio. Simply using these defaults (that is, going with the grain) requires less conscious effort, time, and resources from the developer than going against these defaults. Indeed, the whole reason one might use a commercial game engine in the first place is to have access to these defaults, rather than having to develop everything themselves from the ground up, or having to hire a programmer to do so as discussed in the previous chapter.

The notion that a game engine has a particular look and feel is not new. In the 1990s and 2000s, if a videogame company had paid for a licence to use a particular proprietary engine, this was often viewed as a major selling point. The box for Raven Software's *Star Trek: Voyager— Elite Force* (Raven Software, 2000) proudly proclaims that it is 'powered by Quake III', referring to the id Tech 3 engine, first associated with *Quake III: Arena* (id Software, 1999). While this marketing rhetoric often resembled the 'technobabble' that Dominic Arsenault (2017) highlights as a faux-literacy of the technical aspects of videogames used, primarily, to build consumerist allegiances to particular platforms, it still helped create tangible and perceptual connections between different videogame works. This was especially true for the PC market, where, unlike consoles, videogames were not readily distinguishable by hardware

---

[1] Audio elements are just as crucial as visual elements, and we could just as easily discuss the 'look and sound and feel' here. However, for the sake of brevity, we are restricting our analysis to visual elements. For discussions of the role of the audio in the embodied experience of play, see Keogh (2018) and Collins (2008).

Fig. 4.1   Screenshot of *Doom* (id Software, 1993) (taken by the authors)

platform. A player might not know exactly what it meant to say *Elite Force* was powered by *Quake III: Arena*, but given that association, one could feel certain similarities. Likewise, branding affiliations such as that between Epic's Unreal engine and the *Unreal* franchise of videogames from which its name derives, and the affiliation of the Cryengine engine with studio Crytek and its *Crysis* franchise, work to draw explicit links between videogames that share a game engine provenance.

id's Doom engine, discussed in Chapter 1, provides an instructive example of an engine's grain impacting a videogame's look and feel. id's Doom engine (later rebranded id Tech 1) was used for a range of videogames after the release of *Doom* in 1993, including the fantasy-themed *Heretic* (Raven, 1994) and the sponsored adver-game *Chex Quest* (Digital Café, 1996). While these videogames differ in theme, tone, and atmosphere, they share clear similarities. For example, as illustrated in Figs. 4.1 and 4.2, *Doom* and *Chex Quest* are both first-person videogames where the player navigates 3D polygonal spaces and encounters 2D sprite enemies. In these screenshots, we can see how they both depict the heads-up display (HUD) at the bottom of the screen, with the player's ammo, health, and armour status arranged from left to right.

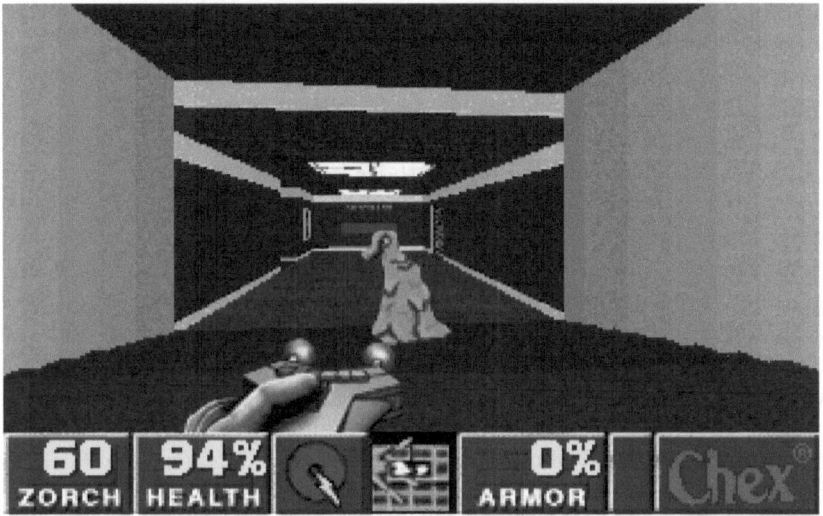

Fig. 4.2 Screenshot of *Chex Quest* (Digital Café, 1996) (taken by the authors)

Both videogames, too, depict the player character's face in the middle of the HUD, with a weapon centred directly above it. While the tone, intended audience, and even genre of these videogames are quite different, in these screenshots one can clearly sense a shared composition and texture. Less explicit but directly contributing to the look and feel of id Tech 1 videogames is how the main menu is navigated, with a cursor token snapping from one menu item to the next, rather than a freely movable mouse cursor. Or the fact that because of how the Doom engine renders its three-dimensional spaces, it only provides a top-down 2D view for the developer to produce levels, and thus, it is technically impossible to place one room directly above or below another, thus influencing what sort of architecture and spaces the player can engage with. Cynically, one might call each of these videogames 'reskins' of *Doom*. More constructively, though, the different videogames made in the Doom engine demonstrate how the game engine provides a base skeleton that gives a fundamental shape to the videogame built up from and around that engine. The videogames in Figs. 4.1 and 4.2 do not look and feel identical, but they definitely look and feel similar through their shared traits, which are derived from the grain of the Doom engine.

While Unity is much more complex and versatile than the Doom engine, it still has a look and feel baked into its default toolset. In an analysis of how Twine was repurposed by marginal—and specifically LGBTIQA+—videogame makers, Alison Harvey (2014: 97–98) demonstrates how software tools that are 'conceptualized as [...] technolog[ies] of game-making' from the ground up have 'presumptions' embedded in their 'structure and paratexts'. As Harvey (2014: 97–98) writes, 'the presumptions about what constitutes a fundamental game design process become clear in the tutorials of many development programs [...] this includes shooting a projectile from one sprite to, or more accurately, at another'. As an example of this, Unity's tutorial video for 'raycasting'—a commonly used function for casting a line through the virtual space from a specific point at a specific angle to see what is in a certain direction—uses the prominent example of a character shooting a gun as a means of demonstrating what this very general method can be useful for. More mundanely, Unity's default 'first-person controller' prefabricated component can be dragged into a scene to create, in a second, all of the elements required to allow the player to walk around a 3D space using typical keyboard and mouse or gamepad controls. This prefab is part of Unity's core code framework and is used commonly by amateur and student developers for first-person shooters and 'walking simulators'.[2] While the first-person controller has a number of variables that can be altered in the inspector pane, it possesses commonly used default settings, which means there is a default way of moving, a default speed, a default field-of-view, and a default footstep sound, the latter of which can often be heard in many Unity-developed first-person videogames.

To stress, these elements can be edited or redesigned on a very low-level. If the designer wishes to create their own custom first-person controller from scratch, they can, and indeed, many do. But as discussed above, such low-level work is hard to justify for many smaller teams lacking the resources and time to build something from the ground up. Indeed, this is probably why they have chosen to use Unity in the first place: to avoid investing unnecessary resources in the development

---

[2] 'Walking simulator' is a (somewhat contentious) term used to describe first-person videogames typically focused on exploration, observation, and narrative, as opposed to shooting, platforming, or puzzle-solving (see Muscat 2018).

of custom solutions and tools. The fact that Unity *can*, theoretically, be manipulated on such a low-level, but often is not, perhaps speaks to the derision with which Unity games are discussed when they 'feel like' a Unity game. While a videogame made in the Doom engine or with Twine might be perceived as adhering somewhat quaint creative constraints that a creator has creatively worked within or around, using the default footstep noises that come with Unity's first-person controller can be perceived as lazy, unprofessional, or derivative.

Indicative of this perspective is former videogame journalist and YouTuber Jim Sterling, who has built a sizeable audience by playing and mocking what he perceives as 'bad' videogames—that is videogames built by small and inexperienced teams, who often rely on using large numbers of prefabricated assets. Inevitably, many of these videogames are made in Unity. In one particular video rallying against Unity for a perceived glut of 'bad' videogames on marketplaces such as Steam, Sterling calls Unity the 'engine of choice for the laziest of developers' (Sterling 2017: n.p.). While Sterling is willing to admit that plenty of 'good' videogames are made in Unity, he claims these are typically developed by studios using the Professional licence, and thus, these 'good' videogames do not have a 'Made in Unity' splash screen. Thus, for Sterling, players are only made aware that a videogame is made in Unity when it is a 'bad' videogame made with the Personal licence. Inherent in such a hypothesis is a direct correlation between budget and quality, and a conflation of 'professional' products with 'amateur' works that will be unpacked further in the next chapter. Important, here, is that it is a particular 'look and feel' of Unity that is explicitly perceived as denoting a videogame of bad quality, and it is the broader range of marginal videogame makers seemingly empowered by Unity's pricing and accessibility that bare the brunt of this criticism and, often, subsequent harassment. Thus, Unity's explicit performance of neutrality directly clashes with the grain its users must inevitably work with and poses a challenge to its claims of democratization. Subsequently, instead of a videogame's particular relation to its engine representing creative expression within constraints, it comes to represent, to a particular audience at least, a perceived *lack* of creative expression, a 'making do' with what is available as opposed to creating a bespoke work. Videogame developers who use Unity must navigate these potential responses and, in the process, decide whether they go 'with' or 'against' the grain of Unity.

DESIGN PRINCIPLES AND DESIGN STANDARDS

The further one strays from 'the path of least resistance' in Unity, the more likely they are to encounter challenges and complications—not only in the design process but also in the methodological and disciplinary assumptions one brings to the design process. Terry Burdak, lead designer of the videogame *Paperbark* (Paper House, 2018), was particularly articulate on these issues. *Paperbark* is a Unity-developed videogame where the player explores an Australian bush environment as a wombat. It has a very distinct watercolour aesthetic that, as Terry explained, is attempting to capture the look and feel of an Australian summer and is deeply influenced by Australian illustrators such as Julie Vivas. Terry is an experienced graphic designer and, before he was a videogame designer, had extensive experience with typesetting and graphic design using a wide range of cultural software.

Not five minutes into the interview, Terry described commercial game engines as 'janky as fuck, because of all this legacy stuff from our computer science history' and highlighted key differences between commercial game engines and Adobe's Creative Suite, noting that 'they're worlds apart, and they shouldn't be, considering they're both meant to be design tools'. Terry suggested that Unity is 'a step in the right direction' as far as making development tools more accessible, but he stressed that 'there's still so much assumed knowledge about computers, about games, about computer science' embedded in the engine. He singled out the visual appearance of the editing interface as an example:

> They've tried to make everything 3D in the engine. You just look at the UI [User Interface] and it's just these 3D things and you rotate this 3D space to be able to do a menu. Why would anyone want to do that? Just like these weird horizon lines, but then the grid's on a 2D plane. It's like, why is the grid on a 2D plane if we're in a 3D space and you want everything to be 3D? It's really confused. I think there's never been any really clear design methods for creating games with a lot of these 3D engines.

Following on from the above, Terry distinguished between design 'standards' and design 'principles'. While Unity has provided developers with something akin to universal design *standards*—largely by virtue of the fact that it has become such a dominant tool—its user interface lays bare the fact that the overall craft of videogame design lacks common design *principles*. As a contrast, Terry offered the example of Adobe Photoshop. In its user interface, Photoshop draws from an extensive repertoire of pre-existing design principles from graphic design, illustration, and photography. In Photoshop, the user can, for example, bring up a digital ruler to measure an image, whereas no such tool exists in Unity. Terry's custom solution for measuring spaces in Unity was to create 3D blocks and manually space them apart in the environment. 'You have to jump through every hoop of measurement', he explained of Unity, 'then you have to make the hoops to jump through to then make more hoops'. Terry's ad hoc measurement solution is a design standard shaped by Unity's affordances, rather than a pre-existing design principle or method that predates videogame development. Historically, the lack of design principles in videogame development is a problem that can be tied to the secrecy and protectionism surrounding programmer-centric pipelines in the 1990s. This is discussed explicitly by O'Donnell (2014) as part of the 'developer's dilemma' stifling innovation and solidarity in the videogame industry. Tellingly, a lack of design principles was a frustration most prominently voiced by participants who had transitioned into videogame development from adjacent creative fields.

In *Software Takes Command*, Manovich (2013: 113) makes a similar distinction between standards and principles through his discussion of 'media-independent' and 'media-specific techniques'. Media-specific techniques are uniquely tied to a specific medium—think Unity's inspector window or its component-oriented design system. Media-independent techniques are, by contrast, concepts that can be implemented to work across a range of different software tools and data types—think cutting, copying, and pasting. Media-independent techniques tend to retain their core principles even as they move between different software environments. As an example, consider 'cinemachine', a virtual camera tool originally made as a plugin for, and then acquired by, Unity. Cinemachine aims to provide developers with accessible tools for composing in-game cutscenes. To this extent, it incorporates several media-independent techniques familiar to anyone with a basic understanding of video editing tools. One such example is the 'timeline', which—much like a timeline editor in any sound or video editing software—enables the developer to make live edits on a cutscene by manipulating video segments, which are represented as 'chunks' in the timeline. Cinemachine also incorporates techniques such as 'track and dolly' (as in tracking shot and camera dolly), which enables the virtual camera to track a target by moving along a path determined by the user. Techniques such as timeline and track and dolly are embedded in a web of Unity-specific cinematic techniques, such as 'deadzone aiming', which refers to a technique where the camera only tracks its target if the target moves outside an adjustable window of sight. This is useful in, for example, 2D platforming games, where it is common for the camera to track the player character only once they move beyond the very centre of the frame. This collage-like composition of techniques is typical of what Manovich (2013: 268) calls 'deep remixability' in software culture, in which 'designers remix not only content from different media but also their fundamental techniques, working methods, and ways of representation and expression'. This deep remixability means that, in most cases, it is very difficult to identify the provenance of a Unity-developed videogame—that is, which of the videogame's various elements (including art, music, physics, and so on) were created using Unity's default tools, which were created in and imported from other tools (although this information may be revealed in splash screens or credits), and which were the result of customized add-ons and plugins. Cultural software are the intersection where these principles, standards, and processes of deep remixability meet.

While Unity builds on many media-independent techniques to make aspects of videogame development more intuitive, it nevertheless relies on a multitude of specialized techniques that, as Terry noted, have been inherited from histories of computer science and programming, and which therefore come across as deeply unintuitive to developers from other disciplinary backgrounds. Moreover, developers need to understand how to utilize these techniques in a way that is legible to the engine; otherwise, they will simply not work. 'The thing I find really hard about games is it's one of the only creative methods that tells you "no"', Terry explained,

> Say you were to pick up a musical instrument and start moving your hands around, making noise. It's not like if you played a scale wrong the thing would then shut down and not let you continue until you've figured out why you did it wrong. And then figure out a method to fix it so then you can start back up the creative process again. But with games you literally get error messages that prevent you from being creative until you fix them.

Aspects of Unity's grain, then, cannot simply be 'worked against' if one wishes to actually compile and run the project. Respondents such as Terry, who were experimenting with unorthodox designs or who occupied something of an alternative or marginal position in videogame culture, were the most likely to voice their frustrations with Unity's editing interface. Having not typically worked in the videogame industry prior to the widespread adoption of commercial game engines, these respondents typically saw Unity as limited compared to other cultural software they were experienced with, as opposed to an improvement in accessible videogame-making software specifically.

We must emphasize, however, that most respondents disagreed with the notion that Unity had inhibited their design methodologies in a meaningful sense. Even respondents such as Terry often qualified their criticisms of Unity by first praising the engine for establishing much-needed design standards in the videogame industry. When so many developers are using the same tools as opposed to being locked inside a proprietary engine's black box, a common language can begin to emerge between developers. Reflecting on this, Morgan made the important point that 'a lot of the people who have had success making games wouldn't have been able to make games without being able to get an off-the-shelf toolset [...] I think the fact that you see games that are built by two, four, five people competing on Steam alongside games that were built by 400 or 500 people [...] is really a testament to what engines offer'. Yet, simply by virtue of its ubiquity and accessibility, Unity's design standards potentially become design principles, as the way to do something in Unity becomes associated with the way to do something in videogame development, period. Developers can either work with the grain of Unity or attempt to work against it, and working with is almost guaranteed to have fewer complications than working against. To quote Chun (2011: 21) once again, 'using free software does not mean escaping from power, but rather engaging it differently'. And thus, we return again to the ways in which Unity's grain is at once democratizing and homogenizing, allowing more people to make more works of a greater diversity and eclecticism of genres, but which nevertheless often look and feel like works created in Unity.

Iterative Design

Unity's ability to both diversify and homogenize, often at the same time, can be understood through a particular aspect of its grain that we articulate here as 'iterative design'. At a number of scales, Unity encourages reuse and repetition with incremental changes. This is visible in the moment-to-moment workflow where the videogame can be played within the editor following every minor tweak, with one component often tweaked a number of times in quick succession. Iterative design is visible, too, in the way a Unity project is 'built up' from individual components and objects, with the overall project broken down into more discrete iterative units of development. It is also visible in the repetition and reuse of standard or popular assets, be they included with Unity as a default package, downloaded from the Asset Store or a website, or distributed and reappropriated through the extensive network of discussion boards and tutorials that surround Unity. At each of these scales, Unity's ability to 'empower' developers is intimately connected to its ability to allow developers to repeat, reuse, and recycle. Iterative design, here, stands in for a radical 'openness' that is in stark contrast to the hegemonic restrictions of proprietary engines.

These very elements, however, stoke negative and anxious discourses around Unity. Figures such as Sterling and the countless threads on videogame discussion boards decrying 'bad Unity games' directly and explicitly blame Unity's grain of iterative design for a large number of low-quality videogames oversaturating digital marketplaces such as Steam. In the same video criticizing Unity already referenced above, Sterling (2017: n.p.) delineates between those 'competent' developers that apparently put in the effort to ensure their Unity games do not look and feel like Unity games, and those 'no-effort chancer[s] who thoughtlessly upload pre-bought maps and fill them with pre-bought characters'. Needless to say, this represents both a gross misunderstanding of the process of videogame development and a particularly narrow and consumerist notion of what might be considered a videogame of 'good' quality. These aspects and their cultural origins will be unpacked further in the following chapter, but here, it is important to consider why Unity's grain of iterative design is so negatively received by some.

When is it preferable to follow a game engine's path of least resistance—to, in other words, go with the grain—and when is it advantageous to fight against it? As discussed in the introduction to this chapter, specialized game engines such as Twine, Bitsy, and Pico-8 tightly restrict what the developer is able to produce by tailoring development towards specific genres or aesthetics, whereas commercial games engines such as Unity and Unreal market themselves as neutral and general purpose. As such, videogames made in Bitsy very much look like 'Bitsy games', and videogames made in Twine very much look like 'Twine games'. Yet, while the expression 'Unity game' is commonly used with derision, an expression like 'Bitsy game' is more commonly used as a genre marker, not dissimilar to 'first-person videogame'. The explicit 'Bitsy-ness' of a Bitsy videogame constitutes an aspect of the creative expression of its creator working with particular tools. Similarly, a Pico-8 videogame might be applauded for its imagining of a particular genre or idea within Pico-8's tight technical limitations. Around these individual cultural software, design principles become popular and normalized common-alities from which individual creations and cultural scenes can be built. Contrast this with Unity's iterative design standards, which are often per-ceived negatively through anxieties surrounding 'asset flipping' and the 'indiepocalypse'. Ironically, the elements of Unity that empower a wider range of developers are the same elements that are seen to be the most generic and 'mass-produced'. The proliferation of 'personal', experimen-tal, and hobbyist videogames is, somewhat contradictorily, often con-flated with a proliferation of impersonal, mass-produced, and soulless videogames.

These tensions have a similar precedent in Walter Benjamin's (1969 [1935]) reflections on the 'mechanical reproduction' of art through industrial processes of production, printing, and broadcasting in the twentieth century. An original artwork, once mechanically reproduced, is expunged of what Benjamin (1969 [1935]: 220) calls its 'aura'—that is, 'its presence in time and space, its unique existence at the place where it happens to be'. By the same token, however, mechanical reproduction emancipates works of art from their anchorage to the dead weight of tradition, ritual, and history. Mechanically reproduced art is, therefore, accessible in a way that traditional art is not—accessible to people who, by virtue of their alienation from bourgeois traditions, rituals, and histo-ries, have not typically enjoyed the privilege of participating in the sphere of artistic production and appreciation. We can apply a similar interpre-tative lens to Unity and the backlash against Unity games. While some Unity games may indeed lack an 'aura' attributed (rightly or wrongly) to videogames developed using custom or specialized engines—that is, videogames that possess a clear and identifiable engine provenance; that communicate something of the time, place, and space in which they were made—they are nonetheless accessible to broader demographics in a way that people steeped in the traditions, rituals, and histories of videogame development may find troubling or threatening. In Chapter 6, we argue that the backlash against Unity games is consonant with a pervasive 'hatred of democracy'—or, in the context of Unity, 'hatred of democratization'—in videogame culture.

Another way to think through these tensions is to consider the notion of craft when it comes to creative software practice. In his critical history of hobbies in North America, Steven Gebler (1999: 156) highlights the handicrafter as evoking 'the mythical purity of the preindustrial artisan' and defines hobbies generally as emerging in the industrial age so as to provide an opportunity for 'fulfilling labour' in the leisure time for those alienated from their work in the Fordist factories. Elsewhere, Susan Luckman (2015) explores the role of craft as a (gendered) creative economy, looking specifically at the re-emergence of handicrafts through platforms such as Etsy. For Luckman (2015: 68–69), '[t]he handmade object is marked by its solid oneness in the world, and is a sign of consumer distinction in a globalised marketplace increasingly marked by a lack of product differentiation: the handmade appeals to people in search of the unique'. While Luckman and Gebler are both talking about physical, handmade objects, it is not a stretch to apply such ways of thinking to the creative work of videogame developers and the cultural reception of that work by videogame players. Here, the iterative design enabled by commercial game engines generally and Unity specifically—arguably the most 'democratic' aspect of Unity—becomes ironically perceived as the least personalized aspect of industrial cultural production. The 'solid oneness' of the Unity game is diluted by the use of seemingly mass-produced, prefabricated components and assets 'flipped' and used again and again. Thus when a developer goes with the grain of Bitsy, they are seen to be producing something handcrafted and personal with the tool, whereas a videogame going with the grain of Unity is perceived as iterative and depersonalized. This can be seen in cultural software more generally, too, such as anxieties that using Photoshop to touch up photographs dilutes the essence of authentic photography, and the typical derision directed at electronic music production compared to music performed with 'real' instruments.

These anxieties point towards a lack of comprehension and literacy regarding the nature and shape of the creative process in the context of cultural software. Through rhetorics of democratization and empowerment and seemingly opening up (but also homogenizing) cultural production for a broader range of potential creators, cultural software seemingly put at risk the oneness, the aura of the cultural work they enable. Such anxieties are not unique to digital environments. Gebler (1999: 260), for instance, references Deborah Nelles (1978) to look at discourses around 'hobby kits' such as Paint By Numbers kits and prefabricated jigsaw kits, which 'deprive hobbyists of an opportunity "to develop their creativity and skill"'. But cultural software specifically and explicitly obscure both the material labour of their users and the grain of the software: the material affordances and learned design principles and imposed standards that such labour is always in conversation with. The next chapter, therefore, turns to the topic of literacies and how Unity is understood by its different constituents—videogame developers, players, and critics.

References

Anable, Aubrey. 2018. *Playing with Feelings: Video Games and Affect.* Minneapolis: University of Minnesota Press.

Arsenault, Dominic. 2017. *Super Power, Spoony Bards, and Silverware: The Super Nintendo Entertainment System.* Cambridge: MIT Press.

Benjamin, Walter. 1969 (1935). "The Work of Art in the Age of Mechanical Reproduction." In *Illuminations: Essays and Reflections,* edited by Hannah Arendt and translated by Harry Zohn. Boston: Houghton Mifflin Harcourt.

Brown, Lisa. 2016. "Vector 2016—The Nuance of Juice Talk." YouTube, September 9. https://www.youtube.com/watch?v=qtgWBUIOjK4.

Bucher, Taina, and Anne Helmond. 2017. "The Affordances of Social Media Platforms." In *SAGE Handbook of Social Media,* edited by Jean Burgess, Thomas Poell, and Alice Marwick, 234–253. Los Angeles: SAGE Publications.

Chun, Wendy Hui Kyong. 2011. *Programmed Visions: Software and Memory.* Cambridge: MIT Press.

Collins, Karen. 2008. *Game Sound.* Cambridge: MIT Press.

Gabler, Kyle, Kyle Gray, Shalin Shodan, and Matt Kucic. 2005. How to Prototype a Game in Under 7 Days. Gamasutra, October 26. http://www.gamasutra.com/view/feature/130848/how_to_prototype_a_game_in_under_7_.php.

Gebler, Steven M. 1999. *Hobbies: Leisure and the Culture of Work in America*. New York: Columbia University Press.

Harvey, Alison. 2014. "'Twine' Revolution: Democratization, Depoliticization, and the Queering of Game Design." *Game* 3: 95–107.

Keogh, Brendan. 2018. *A Play of Bodies: How We Perceive Videogames*. Cambridge: MIT Press.

Luckman, Susan. 2015. *Craft and the Creative Economy*. New York: Palgrave Macmillan.

Manovich, Lev. 2013. *Software Takes Command*. Cambridge: MIT Press.

Muscat, Alexander. 2018. "Ambiguous Worlds: Understanding the Design of First-Person Walker Games." PhD diss., RMIT University. https://research-bank.rmit.edu.au/eserv/rmit:162562/Muscat.pdf.

Nelles, Deborah. 1978. "From Artisan to Courtesan: The Rationalization of Labour and Leisure." Masters thesis, McMaster University.

Norman, Don. 2013. *The Design of Everyday Things: Revised and Extended Edition*. New York: Basic Books.

O'Donnell, Casey. 2014. *Developer's Dilemma*. Cambridge: MIT Press.

Sterling, Jim. 2017. "Unity Has an Image Problem." YouTube, July 24. https://www.youtube.com/watch?v=-z4_bjyJ4EM.

Swink, Steve. 2009. *Game Feel: A Game Designer's Guide to Virtual Sensation*. Burlington: Morgan Kaufmann Publishers.

Wright, Steven. 2018. "There Are Too Many Video Games. What Now?" Polygon, September 28. https://www.polygon.com/2018/9/28/17911372/there-are-too-many-video-games-what-now-indiepocalypse.

CHAPTER 5

Literacy: Articulations of Unity Across Development, Education, and Enthusiast Contexts

Abstract This chapter considers the different ways that audiences invest meanings in a cultural software. Inherent in any of these understandings of the role and mediations of a cultural software are assumptions as to just which skills, knowledges, and aesthetic decisions are fundamental to the process of creating works within any given medium. This chapter thus considers the *literacies* that mediate the perceptions of three different groups towards Unity: videogame developers, tertiary students and educators, and the enthusiast videogame press. Through an overview of these varied perspectives of what it means to use Unity, this chapter provides ways of considering how different literacies influence different understandings, positive and negative, of the role of a cultural software within a cultural field and within society more broadly.

Keywords Unity game engine · Videogame development · Literacy · Education · Enthusiast press · Gamer culture · Asset flips

Cultural software are, to borrow Annemarie Mol's (2002: vii–viii; cf. Banks 2013: 56) term, 'multiple objects' that carry different meanings for different people. For the graphic designer, Adobe's Photoshop represents a crucial tool for streamlining the creation and editing of images. For journalists, Photoshop represents a need to be wary of 'photoshopped' images. For design educators, Photoshop represents a vital skill set employers expect of graduates. As discussed throughout this book, Unity specifically and game engines generally have similarly been imbued with different meanings by different people, who have different understandings of and relationships to game engines. While many developers see the Unity game engine as ultimately empowering, sections of the videogame consumerbase associate Unity with a glut of low quality, impersonal, and mass-produced titles. Researchers and educators, meanwhile, express concern over the supposedly homogenizing effects of game engines. In *Computer Games and the Social Imaginary*, Graeme Kirkpatrick (2013: 105–106) argues that game engines, once introduced in the 1990s, 'rationalized' the 'craft ethos' of videogame development, which had the—'perhaps imperceptible'—effect of closing off possibilities for genuinely imaginative, countercultural, or experimental videogame designs. In a similar vein, Eric Freedman (2018: n.p.) understands game engines as sealing over 'the otherwise latent potential of code' and thus limiting the possibilities for radical sensibilities in videogame development. In this book, meanwhile, our concern with Unity has seen us critique and pay attention to Unity's governance of the developers it seeks to 'empower'.

Inherent in any of these understandings of the role and mediations of a cultural software are assumptions as to just which skills, knowledges, and aesthetic decisions are fundamental to the process of creating works within any given medium. One might, for instance, lament the impact of music software such as Garage Band, if one believes an authentic musician should possess the physical skill of playing a 'real' instrument. In this chapter, we analyse the interplay of these often-conflicting meanings by thinking through the *literacies* that surround cultural software. Here, we understand literacy along the lines of phrases such as 'data literacy', 'systems literacy', or 'information literacy'—that is, 'the development of a complex set of critical skills that allow people to express, explore, question, communicate, and understand the flow of ideas among individuals and groups in quickly changing technological environments' (UNESCO 2006: 150). We are interested in how different people with different literacies invest different meanings in cultural software through their own understandings of how a creative practice is 'supposed' to work and what core qualities the canonical works of a cultural field 'should' possess.

Although the notion that game engines homogenize videogame production makes sense on an abstract level, it does not fully cohere with the understandings of Unity articulated by the developers, students, and educators we spoke with. Perhaps the main objection is that commercial game engines have arguably restored videogame development to something of a folk craft, much like it was prior to the proliferation of proprietary tools (or more accurately, the need to be able to develop proprietary tools yourself) and closed licensing agreements in the 1990s (cf. Keogh 2019). Unity has perhaps 'rationalized' such a craft mythos, as Kirkpatrick (2013: 105–106) suggests, in the sense that more creators take on board the risks of entrepreneurism afforded by Unity's accessibility (as explored in Chapter 3), but it has also arguably afforded a parallel irrationalized craft mythos beyond commodification. The lower barrier of entry has given rise to a diversity of 'everyday game-makers' (Young 2018; Vanderhoef 2019), making videogames for other-than-commercial reasons, whereas before such an endeavour might be seen as too expensive or time-consuming. As one studio director put it to us, if the 'standardization' thesis surrounding commercial game engines was true, then by now we would have witnessed a 'homogenization of the medium and the culture, whereas it's been the opposite—we've seen a vast range of diverse voices and play experiences and types of games to where some people even argue whether something is a game or isn't a game'.

Unity does not simply 'make it easier' to make videogames. Rather, it automates and offsets previously fundamental skill sets (namely software development) that were once utterly required to undertake videogame development. Unity makes it possible for people with *different* skill sets and technical literacies to make, share, and evaluate videogames, and thus, Unity videogames are both more homogenous and more diverse in *different* ways than videogames of previous decades. What does it mean, then, to assume that Unity is responsible for a dearth of skills crucial for innovation and experimentation, when the opposite seems to be true? What does it mean for Unity to try to explicitly 'empower' developers by allowing the circumvention of those skills? This chapter answers these questions by examining three different constituencies of Unity's circuits of cultural software. The first section explores why the videogame-makers we interviewed use Unity. For many, Unity is quite simply the engine they already know, and using it is hardly a conscious choice at all. Their enrolment into Unity's governance circuit was often unconscious. The second section considers the role of videogame development education programs, wherein commercial game engines such as Unity have seen a significant uptake. Here, we consider what it means for Unity to not just be voluntarily chosen, but prescribed at an institutional level. The third section returns to the enthusiast literacies and *faux-literacies* that lead some players and critics to dismiss Unity-made videogames out of hand. To stress, these are not the only constituents whose varying literacies of Unity are important, and an entire chapter could easily be spent on each of these constituents. Yet, through a brief overview of these varied perspectives of what it means to use Unity, this chapter provides ways of considering how different literacies influence different understandings, positive and negative, of the role of a cultural software within a cultural field and within society more broadly.

'WHY DO YOU USE UNITY?'

There are multiple questions to consider when it comes to selecting a game engine for a videogame project or series of projects. As already discussed, engines such as Unity and Unreal provide general-purpose toolsets, whereas engines such as Twine and Pico-8 provide much more specialized affordances. Some engines grant access to source code, while others remain black-boxed at a low-level. Some presume access to other skills and software, such as 3D modelling or audio composition, while others restrict the user to the engine's built-in tools. There are various licensing agreements and subscription models to contend with, not to mention proprietary considerations for certain engines. Some engines support a range of programming languages (the open-source game engine Godot supports C#, C++, and its own GDScript), while others support only a single language (Pico-8, for example, only supports Lua). There is also the question of how an engine will fit into an existing development pipeline, as discussed in Chapter 2. How, then, do developers arrive at a decision to use one engine in place of another? What existing skills, competencies, and understandings—in short, what literacies—influence such a decision?

Many hobbyist and student respondents were actively experimenting with different technologies, techniques, and methodologies at the time of being interviewed. For example, Erika Verkaaik, an independent videogame-maker, educator, and Masters student based in Brisbane, explained that they often have a specific idea for a videogame before choosing an engine that will enable them to most effectively and efficiently develop that idea. For example, if they have an idea for a text adventure videogame, they will use Twine, whereas if they have an idea for a 3D videogame, they will usually opt for Unity. Erika also explained that they are often inspired to use a particular engine after they play a particularly creative or unorthodox videogame built in that engine. Hobbyists often identified an engine's accessibility as a key priority, referring to the intuitiveness of the interface, the ability to switch rapidly between editing and prototyping, the amount of coding required to adequately realize a creative vision, and the cost of obtaining a licence. Several hobbyists and students preferenced Unreal on the basis of that engine's Blueprint visual scripting system, which allows for programming without the need to know a deep amount of syntax. Nonetheless, although respondents in this category often used a variety of engines, most considered Unity to be their primary engine—unsurprising, considering the report cited in this book's introduction that over 75% of Australian studios primarily use Unity. In our interviews, Unity's popularity often came down to its status as a general-purpose engine, its 'free' or low-cost availability, and its robust support networks—both online and through local communities of Unity-using peers.

For professional and studio-based respondents, engine preference was driven by more pragmatic concerns. For Trent, the decision to use Unity was based on the fact that his studio was originally composed of geographically dispersed developers and coders with proficiency in C#, meaning that 'anyone could download Unity and immediately jump into it'. Morgan, who in Chapter 3 described his Brisbane-based studio Defiant Development as 'a games company, not a technology company', discussed his studio's decision-making process in the following way:

> The incentives that had previously existed in Australia around building your own technology and doing your own R&D no longer existed, so there wasn't a financial incentive to be building technology instead of games. Building games was what we wanted to do. We didn't want to spend a year laying the groundwork for the game so that we could get it out. So we started using Unity, which let us build things quickly and get them out to market. That's turned out to work out really well, and we haven't seen a need to move off it.[1]

---

[1] See Banks (2013) for a case study of the 'incentives' that existed for Australian videogame companies to build custom tools in the late 1990s and early 2000s. Banks offers a detailed ANT-inspired account of the Brisbane-based studio Auran (now N3V Games) from 1998 to 2000, when the company was investing a huge amount of resources in the development of a proprietary engine known as SAGE, later renamed Jet. The engine was to be utilized not only for in-house development purposes, but also as a means of attracting

Morgan went on to explain that, since adopting Unity, Defiant Development has worked on a range of projects, from mobile videogames to augmented reality projects, and that 'Unity has handled that transition pretty well'. House House, a four-person team introduced in Chapter 3, switched to Unity after first building a custom web engine for what started out as a hobbyist project. House House's decision to switch to Unity was, according to designer Jacob Strasser, 'always a given'. The team already had a level of familiarity with Unity after having ported their earlier web videogame to the engine in order to publish it on multiple platforms. The main alternative for House House—the Unreal engine—'seemed harder to learn of the two, and we were absolute beginners'. Jacob's colleague Michael McMaster spoke directly to the power of Unity's network effects when he added that 'we've got a bunch of friends that we can ask for help if we have issues [with Unity], which isn't necessarily true for Unreal'.

There were some instances where a developer's identification with (or against) a particular game engine (or cultural scene in the development community) appeared to be a key factor shaping their engine preference. For example, one developer described his preference for Unity—and his distaste of the alternatives—in the following way:

> It's really weird but I don't want to use Unreal because I don't like the name. I don't like the way it looks. I just think it's ugly. [The engine] GameMaker is also ugly [...] Unity is probably the most inoffensive looking game engine, so that's a big reason why we probably still use it [...] Unreal feels like daggers in you or something, you know what I mean? It's got like, horns or something. I don't know. Hate it. I hate it. It's like Alienware and all that other stuff [...] It's really gross.

A game engine can intersect with a developer's subject position, not just in terms of their preference for a particular editing interface or licensing structure, but also in terms of their cultural positioning, politics, taste, gender, and identity. We found that game engines such as

---

external licensees. Interestingly, Banks (2013: 51) observes that Auran set itself the impossible task of developing an engine that could 'do everything', which, according to one of his interviewees, was 'all about marketing [and] hype'. This speaks to the long-standing fantasy of a universal game engine, discussed in Chapter 2.

Unreal are often associated with 'gamer' subject positions (see Shaw 2012), which could be due to any number of articulations, such as the articulation of the Unreal engine to its origins in the Unreal series of first-person shooter videogames. The notion that a developer's subject position can overlap with their engine preference is particularly visible across grassroots engines such as Twine and Bitsy, which have come to be associated with fringe and gender-diverse videogame-making communities (Harvey 2014; kopas 2015). In a different context, Yoyo Games' Gamemaker Studio, which was once disparaged in a manner not dissimilar to the backlash against 'Unity games', has come to be associated with an 'indie' coolness. Developers such as Cactus (responsible for videogames such as *Hotline Miami* [Dennaton, 2012]) and Vlambeer (responsible for videogames such as *Nuclear Throne* [Vlambeer, 2015] and *Super Crate Box* [Vlambeer, 2010]) proudly label their videogames as 'made in Gamemaker'. Unity, however, was rarely discussed by respondents in relation to personal identity—which is not to say that personal identity did not play a role in their choice of Unity, but rather that Unity has become something of a 'default' option.

While reasons for choosing Unity were often more pragmatically about access to resources or skills, many studio-based respondents struggled to coherently recall the processes that went into their decision to use Unity in the first place. In fact, when we posed this question in interviews, it was often our impression that this was the first time that team members had ruminated on why they had, in fact, chosen to use Unity over, say, Unreal. This could be explained by the fact that the decision to use a particular engine is often made by people in senior management and software engineering positions—people not always present in the interviews. But more often, our sense was that the decision to use Unity was, to borrow Jacob's words, simply 'a given' in most scenarios.

A minority of respondents had decided to forego Unity or any other commercial game engine to instead create their own custom engine. These respondents possessed a clear passion for tool development (as opposed to simply videogame development) and were typically developing niche, high-performance, or unorthodox projects that, in their view, necessitated custom tools. Discussing their desire to produce their own custom tools, these respondents often echoed concerns in the introduction to this chapter that game engines pave over 'the otherwise latent potential of code' (Freedman 2018: n.p.). Lawrence Millar-Madigan, a software engineer working part-time at The Arcade, was independently developing a custom engine when we met him in October of 2018. He described his motivation for doing so as follows:

> Our experience—you and I, what we experience day-to-day—we're not electrons, we're not matter. We're in the ether. Our experience is not a physical one, really. But it's based on extremely complicated and weird physical interactions that are going on. When you make a game from first principles, it's like you can see the way life connects to its own underpinnings. You feel connected to that building of something transcendent, from building blocks that are integral and down to earth [...] That's what motivates me to do what I'm doing.

Just as many studio developers often struggled to recall the reasons why they originally decided to use one engine over another, Lawrence explained that his interview with us was one of the few moments where he felt prompted to distil his thoughts on why he had chosen to develop his own engine. 'You've put me in a spot where I have to express it properly', he explained, 'and I never really expressed it quite like that before. I understand it on an emotional level, but at times I wonder, "what am I actually doing?"'.

We assumed that developers of custom game engines would have a clearer sense of why they had chosen to 'roll with their own tools', but this was not always the case. Quite simply, for many developers, what engine to use (or create) is not a 'choice' at all, but is shaped by a range of factors including identity, community, available resources, and pre-existing skills. Even in large studio contexts with proprietary engines, the engine used for a videogame can simply come down to what is available. For example, in his in-depth journalistic investigation of the beginnings of Respawn Studio and the development of *Titanfall* (Respawn, 2014), Geoff Keighley (2014) highlights how the team spent many months to make the videogame work in Insomniac's Luna engine (originally used for *Ratchet and Clank: A Crack in Time* [Insomniac, 2009]) simply because Insomniac's CEO offered the engine to the young studio for free. Ultimately, however, the videogame was rebuilt in Valve's ageing Source engine, because Luma introduced too many workflow challenges. Given that even larger studios encounter these kinds of problems, it is not surprising that so many small-to-medium-sized studios default to Unity, an engine that prides itself on providing a generalist framework upon which 'any' videogame can be developed. Quite simply, developers often use Unity because they already know how to use Unity, and developers already know how to use Unity because Unity has successfully ensconced itself within local communities as the default option.

UNITY IN TERTIARY EDUCATION

When asked how they first got into creating videogames, most developers' background stories fit one of two common narratives. Either they first encountered videogame development by tinkering with freely available tools (typically Unity, Flash, Twine, or modding, depending on when and how the developer was introduced to videogame-making), or they were first introduced to videogame development when they commenced a tertiary-level videogame development education. How Unity fits into the former of these has already been hinted at in previous chapters' discussions of Unity's self-positioning as a 'democratizing' tool, its low barrier to entry, and its extensive online networks and resources of support. However, formal videogame development education is an equally crucial site in which the particular literacies of Unity are normalised. Increasingly, Unity either explicitly or implicitly becomes the default learning environment for student videogame developers, and schools become a crucial juncture where Unity's dominance is reinforced and perpetuated—not dissimilar to the dominance of Photoshop in graphic design curriculums.

Since the turn of the century, tertiary videogame development programs have emerged globally to teach videogame development from a range of different perspectives, be that art school, computer science, or the creative industries. The proliferation of these schools has been driven, in part, by rapid industry growth in specific regions and by a broader enthusiasm among younger generations to pursue a career in videogame development. While little scholarly attention has been paid to videogame development programs (see Harvey 2019; and Zagal and Bruckman 2008 as notable exceptions), the role and effectiveness of videogame development programs is a hotly and perpetually debated topic among videogame development professionals, and intersects with broader debates around the role of cultural skills education in the contemporary university (Bridgstock and Cunningham 2016). As formal videogame development education is a relatively recent phenomenon, many industry veterans who did not enter the industry via such programs themselves are sceptical of the value of such a pathway, and they often suspect the schools of directly exploiting the dreams and passions of their student cohorts (Warner 2018; Wright 2018). Other developers and educators, such as Robert Yang (2018), are critical of the industry's scepticism towards videogame development education, noting that education programs do not simply exist to funnel students into an industry, and that the industry is naive if it expects graduates to be 'job ready' right out of school, rather than requiring further mentorship, support, and onboarding from their potential employers.

One site of contention for videogame development programs, especially in the early-to-mid 2000s, was the lack of access to proprietary game engines at most videogame education institutions. While some schools were offered access to an engine by a local studio (in part due to that studio hoping to develop a skill pool of potential employees), very few students of this time had access to the tools or platforms of videogame development actually used within the formal industry. Instead, schools often had to make do with either hobbyist tools (such as Flash or Gamemaker), or modding or level editing tools, such as the *Starcraft* (Blizzard, 1998) level editor. Videogame development education during this time was either highly technical (training the programmers who might one day develop a proprietary engine or custom tools), or highly theoretical (training the videogame designers in theories of play and design but with little access to the actual software or practices of the industry).

The emergence of Unity and its subsequent proliferation through the Personal licence (with engines such as Unreal following suit) provides an unprecedented opportunity. Now, videogame development degrees are able to teach the same software environment that is used by the studios their students will potentially be seeking employment at. In Australia in particular, the fact that much of the local industry is already using Unity places pressure on students to have that skill. Further, due to the licensing freedoms of the Personal licence, students are able to distribute, and perhaps even sell, their own student projects produced in Unity (including Asset Store projects)—an opportunity that was less feasible when using hobbyist tools or proprietary level editors. In a number of schools we conducted interviews at, this manifested as an entrepreneurial approach to videogame development education where students were being trained less for future employment, and more for starting their own studios and producing their own videogames. While the accessibility of commercial game engines affords this pedagogical approach, in Australia, where jobs simply do not exist for all graduates, the 'involuntary entrepreneurism' (Oakley 2014) of 'going indie' is seen as both necessary and appealing. In this way, the introduction of commercial game engines to tertiary spaces and the fostering of individualized workflows they afford (see Chapter 3) are not only empowering students, but also reconfiguring how the process of videogame-making is understood and perceived by students.

Unsurprisingly, like other cultural software, Unity Technologies views schools as primary sites for the enrolment of new constituents into their software ecosystem. As illustrated in the previous section, if a student learns Unity, one expects it is Unity that then becomes the default choice for that student's future videogame projects. While Unity once offered an explicit education licence that included a permanent 'for educational use only' waterstamp in the corner of published content, Unity's 'for students' page now redirects students to the same Personal licence as is used by many hobbyists and commercial developers. Educators, meanwhile, are directed towards a repository of resources suggesting ways to teach Unity in the classroom. For several years, Unity also ran a student ambassador program wherein students from different Universities were elected to be Unity champions within their schools. We spoke with one student ambassador who described the role as 'not quite marketing, but like they [Unity Technologies] were trying to catch students at an early stage, and get them using it [the Unity game engine], so then they're "Unity for life"'.

Educators often voiced concerns about students becoming 'just' Unity developers. There was typically a sense that only being able to use Unity was insufficient for being proficient in either videogame design or videogame programming. Instead, it was seen as crucial that students were encouraged to adapt to a range of tools and environments, so they could, as one educator put it, 'learn how to learn an engine'.[2]

---

[2] A representative of a technical college in the Netherlands, who was interviewed for an adjacent project by the authors, discussed taking the opposite approach. They had renamed their degree from 'videogame development' to 'Unity development'. While the degree program still largely focused on videogame development skills, the school representative we spoke to noted that they had identified more extensive employment opportunities for Unity developers, beyond the videogame industry. According to them, other industries, including mining, advertising, the military, and freight shipping, were increasingly looking for developers specifically skilled in Unity. There are some statistics to back up this claim, such as Linkedin's 2017 'U.S. Emerging Jobs Report', which situates 'Unity developer' as the 7th most in-demand job title among US employers (Economic Graph Team 2017).

The further anxiety here is that if students only learn how to use Unity, then they may become perpetually stuck within Unity's ecosystem (see also Deterding and O'Donnell 2016: n.p.)—in the terms of the previous chapter, students risk becoming competent in Unity's design standards, but not videogame development's design principles. For example, Grace, whose videogame *The Haunted Island* we discussed in the opening anecdote to this book, described receiving criticism from her teachers for creating projects that too closely resembled 'Unity games'. This is despite the fact that Grace deliberately and explicitly embraces the grain of Unity as a design decision. We also spoke with a software engineer who was interviewing several University graduates for a position as technical artist in his studio. 'I've interviewed a lot of people now', he explained to us, 'and a lot of them—most of them, in fact—say "I love to know what's going on. I love to understand what's happening". And as soon as I probe a little bit deeper, I'll realize that they haven't actually gone to any lengths to understand what's going on'. In his view, most people are 'happy to leave Unity as a black box', which he found troubling. Other employers, however, saw Unity's proliferation in schools as an unequivocal benefit, as graduates could be more easily assimilated into the Unity-inflected workflows of their studio.

In some of the schools whose educators we spoke to, Unity was mandated by the course curriculum. Many others, however, worked to remain 'engine agnostic' in their curriculum, teaching transferable processes and concepts and expecting students to spend their own time gaining familiarity with a game engine. Cherie Davidson described her role as an educator as assisting '*self*-guidance' and '*self*-teaching':

> I spend most of my time talking to [students] about what they want to do. Walking them through how they think they could achieve it with the engine. More than teaching the 'how to do it', it was more like 'how can you find the community and the APIs [Application Programming Interfaces] that already exist? How can you find that information yourself?'

This approach, however, is only possible due to the extensive networks and resources available around commercial game engines such as Unity and Unreal. Indeed, even when Unity is not required by the curriculum, it still commonly becomes the default due to its sheer ubiquity—it is the engine that other students on team projects are most likely to have experience in, and it is the engine that educators are most likely to have prior experience with.

Thus, ensuring students 'learn how to learn an engine' both future-proofs students against Unity's future business decisions and unpredictable platform redesigns, and also facilitates an entrepreneurial and self-governing approach to videogame development that directly aligns with and is easily absorbed by Unity's own circuit of governance. Students (and by extension, developers) become responsible for their own skill education—required to 'self-guide' and 'self-teach'—in lieu of institutional support structures of upskilling, mentorship, and explicit promotional pathways. In the videogame development curriculum, the commercial game engine discursively strengthens an ideology of creative professionalism that 'is now understood in terms of entrepreneurism' and which reconfigures creative practice pedagogies (including those of the school) as 'the economization of imagination, the marketization of creativity' (McRobbie 2016: 76). In the videogame school, mandated or not, Unity instils literacies of self-governance that reinforce Unity as the default option, as not really a choice at all.

UNITY IN THE ENTHUSIAST DISCOURSE

Throughout this book, we have made reference to anxieties surrounding Unity's bid to democratize videogame development. These anxieties are most dominantly visible within core gamer discourses—in, for example, Jim Sterling's videos—but, as the next chapter will discuss, they also impact professional development communities and digital storefront policies. For gamers, the concern loosely centres on issues of authenticity, and the policing of a cultural field once perceived as subcultural but that is now considered more accessible and diverse. For developers, the concern is that a high quantity of low-quality videogames on digital storefronts will lower the discoverability of their own videogames, as well as erode consumer trust in the videogame industry generally. For those external to core gamer and videogame industry discourses, both concerns might seem somewhat overblown and counterintuitive. While one should of course be critical of any one cultural software company monopolizing a cultural field (as we have attempted to be in this book), surely more people having access to the means to both create and produce videogames can only create more diversity and choice for players? Furthermore, who gets to decide that *all* videogames produced beyond a traditional studio model or without a commercial amount of polish are of 'poor' quality in the first place? Surely, too, the 'indiepocalypse' that developers fear—the notion that there are 'too many games and too many developers' (Wright 2018: n.p.)—is, for better or worse, a reality to be expected for any aspiring cultural worker in any field of cultural production? The idea that there are 'too many bands', for instance, would seem absurd. Indeed, many of the anxieties surrounding Unity seem to be symptomatic of the fact that videogame development is becoming more like any other field of cultural production. Yet, as Pierre Bourdieu (1983: 323) notes, one of the key constituting tensions in any field of cultural production is the struggle over who gets to define just what the field is. We are seeing this play out vividly in the field of videogame production. As a wider range of people gain access to the ability to both distribute and develop videogame works—in no small part through Unity's 'democratizing' effects—new genres emerge, and this leads to debates as to what counts as a 'real' videogame at all.[3] As such, we end this chapter with a consideration of the broader cultural literacies that orbit Unity and the process of videogame development more generally. While such literacies do not necessarily have a direct impact on who is able to use Unity, they are crucial to comprehend how Unity's circuits of cultural software are navigated by a wider range of constituents and how works produced with this software are potentially perceived and evaluated.

This requires us to understand where dominant conceptualizations of the videogame field originated. The mid-1980s brought in a period of 'aggressive formalisation' to the videogame industry that has more recently been supplanted by a period of 'intense in/formalisation' (Keogh 2019). 'Aggressive formalisation' here refers to the period of time in which proprietary game engines and restrictively curated distribution platforms cultivated a very specific image of what videogames 'are' and who they are for. A range of researchers looking at popular videogame discourses of this time have highlighted how videogame publishers and the enthusiast press (often hand-in-hand) manufactured an imaginary 'gamer' identity interested primarily in systemic challenges, technological spectacle, and a large quantity of 'content' (Shaw 2012; Kirkpatrick 2015; Arsenault 2017; Cote 2018; Nicoll 2019). However, as the normalization of digital distribution and accessible commercial game engines took hold in the 2000s, broader fields of both creators and

---

[3] See Harvey (2014: 100) for a discussion of this issue in relation to 'the legitimacy of twine games' created primarily by LGBTIQA+videogame-makers.

audiences have opened up beyond the traditionally hegemonic and masculine 'gamer' cultures. It is this shift to intense in/formalisation—where clear distinctions between 'amateur/professional', 'artist/employee', and even 'player/developer' have crumbled away—where a number of cultural and political conflicts have emerged in videogame discourse. This was most notable in the misogynistic harassment campaign 'gamergate', which coalesced around gamer anxieties directed at videogames created by both LGBTIQA+ and gender-diverse videogame developers (see Shaw and Chess 2015).

Most directly relevant to audience literacies around Unity is how this period of aggressive formalisation instilled particularly technologically deterministic conceptualizations of the process of videogame development. As several firms including Nintendo, Sony, and Sega competed for market dominance through the 1990s, the 'console wars' became defined by (perceived or actual) technological innovations, such that the wheel of videogame history was itself conceptualised through the metaphor of a series of self-contained technological 'advancements' or 'generations' (Therrien and Picard 2016). As a result, the quality of a videogame became intrinsically linked to its technological foundations—whether that be a game engine or console. For example, the UK's *Official Nintendo Magazine* compared the three-dimensional *Castlevania 64* (Konami Kobe, 1999) for the Nintendo 64 with *Castlevania: Symphony of the Night* (Konami, 1997) for Sony's competing PlayStation, noting that 'the N64's miles ahead of the 32-bit competition [...] Whereas the N64 version features fully 3D characters and backgrounds, along with dazzling lighting effects, the PlayStation title is a flat, 2D platform game where the tiny hero has to fight big, cheesy monsters. No comparison, really...' (Clays 1997: 4). Here, 3D innovations are perceived as inherently better than the seemingly old-fashioned and technologically unimpressive 2D art. By a technologically deterministic metric, the videogame being released tomorrow will always be better than the videogame that was released yesterday, and an enthusiastic consumerbase is held in perpetual anticipation for 'the next big thing'.

In his platform study of Nintendo's Super Nintendo Entertainment System, Dominic Arsenault (2017) looks at the marketing discourses used by Nintendo specifically to frame the technologies underpinning its videogames, drawing a distinction between 'technoliteracy' and 'technobabble'. While technoliteracy provides broader audiences with 'access and proficiency in understanding the complexities of technology', popular videogame discourses of this period instead '[used] techno-babble to smash [readers] with complexity and leave them dazzled or beaten senseless' (Arsenault 2017: 80). As it formalised throughout the 1990s, videogame discourse was defined not by a simple ignorance of the complexities of videogame technology, nor solely by a straight-forward technological determinism, but rather by a 'brandishing of fac-tual information and data without context' (Arsenault 2017: 77). This smokescreen of technobabble cultivated particular *mis*understandings of how videogame technologies work and how they intersect with the vid-eogame development process. Technobabble 'insidiously instill[ed] the impression of knowledge in people, hence bringing them into double ignorance—not knowing that they don't know' (Arsenault 2017: 80). Despite videogame development and play cultures shifting dramatically since the 1990s, the above ways of talking about, thinking about, and evaluating videogames continue to linger in the cultural discourses that surround videogame production. One particularly revealing contempo-rary example was provided in 2013, when Sony representatives demon-strated *Killzone: Shadow Fall* (Guerilla, 2013) on *The Tonight Show Starring Jimmy Fallon*. The representative can be heard claiming that the new console's 'eight gigabytes of the fastest unified memory [...] gives us space to develop these characters that you really care about' (Hecker 2013: n.p.). Here, the available technology (provided without context) is directly linked to the ability of developers to craft higher quality narrative content.

If one requires 'eight gigabytes of the fastest unified memory' in order to 'develop these characters that you really care about', then what does that imply about the quality of videogames made in commercial game engines such as Unity—engines that are readily available and much less technologically exciting? Technobabble, as deployed by blockbuster studios and console manufacturers—those who most benefit from tech-nologically deterministic evaluations of videogame quality—instils a par-ticular (il)literacy of the videogame development process among players and critics. Consequently, now that more people more visibly make vide-ogames in/formally—where the technology is rarely the most impressive aspect of the project—their videogames seem technologically homoge-neous even if they are diverse from a creative, thematic, or design per-spective. Technobabble perpetuates a dominant centre and subordinate margin of videogame development where those who continue to have access to proprietary engines and programmer-oriented pipelines are able to produce videogame works that are most readily perceived as being of good quality and demonstrating good craftspersonship, while the broader ecology of videogame developers 'empowered' by commercial game engines such as Unity remain at the fringes, with their own craftspersonship often dismissed as the simple reuse of copy-and-paste prefabricated assets.

Thus, we can now understand how commentators such as Jim Sterling are able to write off developers that take advantage of Unity's component-oriented development environment and affordance of iterative design by using pre-existing assets in lieu of the resources to produce them from scratch as 'lazy' or 'no-bit chancers'. What might be impressive about these videogames—their design, their narratives, their experimentation with the common styles of videogame play—is overshadowed by the common technology that they are enabled by. Following a period of over a decade of tightly controlled and formalised videogame production, a dramatic increase of amateur, marginal, experimental, and derivative works comes to be seen as a direct challenge to the naturalized hierarchies of the field of videogame production. Different literacies around the role and context of videogame development tools, including game engines, directly impact how those tools—and the works produced with them—circulate and are perceived. This puts into context why commercial videogame developers using Unity feel pressured, as highlighted in Chapter 4, to 'go against the grain' and to hide the 'look and feel' of Unity. More broadly, the different ways that Unity is understood by its users, educators, and students, as well as by a dominant gamer demographic, point towards the importance of understanding how and why different constituents become enrolled in the ecosystem of a particular circuit of cultural software. The varying literacies (as opposed to competencies) held by these different constituents form specific ways of knowing, feeling, and identifying, which in turn shape the identity of the software in question as well as constituents' attitudes towards it.

REFERENCES

Arsenault, Dominic. 2017. *Super Power, Spoony Bards, and Silverware: The Super Nintendo Entertainment System*. Cambridge: MIT Press.
Banks, John. 2013. *Co-creating Videogames*. New York: Bloomsbury.
Bourdieu, Pierre. 1983. "The Field of Cultural Production, or: The Economic World Reversed". *Poetics* 12: 311–356.

Bridgstock, Ruth, and Stuart Cunningham. 2016. "Creative Labour and Graduate Outcomes: Implications for Higher Education and Cultural Policy." *International Journal of Cultural Policy* 22 (1): 10–26.

Clays, Simon. 1997. "Return of the Vampire!" *Nintendo Official Magazine* 57: 4–5.

Cote, Amanda C. 2018. "Writing 'Gamers': The Gendered Construction of Gamer Identity in *Nintendo Power* (1994–1999)." *Games and Culture* 13 (5): 479–503.

Deterding, Sebastian, and Casey O'Donnell. 2016. "Game Engines in Game Education: Thinking Inside the Toolbox?" GDC Vault. https://www.gdcvault.com/play/1023034/Game-Engines-in-Game-Education.

Economic Graph Team. 2017. "LinkedIn's 2017 U.S. Emerging Jobs Report." Linkedin, December 7. https://economicgraph.linkedin.com/research/LinkedIns-2017-US-Emerging-Jobs-Report.

Freedman, Eric. 2018. "Engineering Queerness in the Game Development Pipeline." *Game Studies* 18 (3). http://gamestudies.org/1803/articles/ericfreedman.

Harvey, Alison. 2014. "'Twine' Revolution: Democratization, Depoliticization, and the Queering of Game Design." *Game* 3: 95–107.

Harvey, Alison. 2019. "Becoming Gamesworkers: Diversity, Higher Education, and the Future of the Game Industry." *Television & New Media* (OnlineFirst): 1–11.

Hecker, Chris. 2013. "Fair Use." Youtube, March 29. https://www.youtube.com/watch?v=kXnoW2SvQrQ.

Keighley, Geoff. 2014. "The Final Hours of *Titanfall*." http://www.finalhoursoftitanfall.com/.

Keogh, Brendan. 2019. "From Aggressively Formalised to Intensely In/Formalised: Accounting for a Wider Range of Videogame Development Practices." *Creative Industries Journal* 12 (1): 14–33.

Kirkpatrick, Graeme. 2013. *Computer Games and the Social Imaginary*. Cambridge: Polity Press.

Kirkpatrick, Graeme. 2015. *The Formation of Gaming Culture: UK Gaming Magazines, 1981–1995*. New York: Palgrave Macmillan.

kopas, merritt. 2015. *Videogames for Humans: Twine Authors in Conversation*. New York: Instar Books.

McRobbie, Angela. 2016. *Be Creative: Making a Living in the New Culture Industries*. Cambridge: Polity Press.

Mol, Annemarie. 2002. *The Body Multiple: Ontology in Medical Practice*. Durham: Duke University Press.

Nicoll, Benjamin. 2019 (forthcoming). *Minor Platforms in Videogame History*. Amsterdam, the Netherlands: Amsterdam University Press.

Oakley, Kate. 2014. "Good Work? Rethinking Cultural Entrepreneurship." In *Handbook of Management and Creativity*, edited by Chris Bilton and Stephen Cummings, 145–159. Cheltenham: Edward Elgar.

Shaw, Adrienne. 2012. "Do You Identify as a Gamer? Gender, Race, Sexuality, and Gamer Identity." *New Media Society* 14 (1): 28–44.

Shaw, Adrienne, and Shira Chess. 2015. "Reflections on the Casual Games Market in a Post-GamerGate World." In *Social, Casual, and Mobile Videogames: The Changing Gaming Landscape*, edited by Tama Leaver and Michele Wilson, 277–289. New York: Bloomsbury.

Therrien, Carl, and Martin Picard. 2016. "Enter the Bit Wars: A Study of Video Game Marketing and Platform Crafting in the Wake of the TurboGrafx-16 Launch." *New Media & Society* 18 (10): 2323–2339.

UNESCO. 2006. *Education for All Global Monitoring Report: Literacy for Life*. Paris: United Nations Educational Scientific and Cultural Organization.

Vanderhoef, John. 2019 (forthcoming). *Passion, Pixels, and Profit: The New Creative Economy of Indie Game Production*. Ann Arbor: University of Michigan Press.

Warner, John. 2018. "It's Time We Stopped Encouraging Indies." Gamesindustry. biz, October 2. https://www.gamesindustry.biz/articles/2018-10-02-its-time-we-stopped-encouraging-indies.

Wright, Steven. 2018. "There Are Too Many Video Games. What Now?" Polygon, September 28. https://www.polygon.com/2018/9/28/17911372/there-are-too-many-video-games-what-now-indiepocalypse.

Yang, Robert. 2018. "What Is the Game University For?" Radiator, May 4. https://www.blog.radiator.debacle.us/2018/05/what-is-game-university-for.html.

Young, Christopher J. 2018. "Game Changers: Everyday Gamemakers and the Development of the Video Game Industry." PhD diss., University of Toronto.

Zagal, José P., and Amy Bruckman. 2008. "Novices, Gamers, and Scholars: Exploring the Challenges of Teaching About Games." *Game Studies* 8 (2).

CHAPTER 6

Governance: Unity's Democratization *Dispositif*

Abstract This chapter treats Unity's slogan—'democratizing videogame development'—as a policy discourse and *dispositif* that functions to legitimize neoliberal modes of work and identity formation. It discusses videogame culture's conflicted view of democratized development tools by looking at the anxieties surrounding 'asset flipping' and the 'indiepocalypse', both of which are often linked to Unity's 'democratizing' effects. It argues that people feel empowered by Unity not only because of the tools it provides, but also because it appears to make a sustained, policy-driven commitment to democracy and equality in a political environment where such a commitment is typically felt to be lacking. It concludes by discussing alternative 'grassroots' game engines and by pointing to further uses of the circuits of cultural software.

Keywords Software culture · Platform governance · Asset flips · Indiepocalypse · Democratization of videogame development · Unity game engine

It is often argued that today's platform and software companies are supplanting many of the roles, services, and infrastructures once associated with public-oriented governance structures (Chun 2011; Gillespie 2017b; Plantin et al. 2018). Platforms such as Facebook and WeChat are providing 'free' self-management tools in exchange for massive amounts of personal data and, in the process, are becoming increasingly embroiled

© The Author(s) 2019
B. Nicoll and B. Keogh, *The Unity Game Engine and the Circuits of Cultural Software*, https://doi.org/10.1007/978-3-030-25012-6_6

101

in democratic processes and political controversies. Likewise, by providing media creatives with tools that promote self-sovereignty and self-entrepreneurship, cultural software such as Unity and Photoshop strategically position themselves to fill a void left by the erosion of state welfare and cultural policy in Western countries. These developments are having dramatic consequences for traditional ideas of democracy and governance. Legislators are facing pressure to regulate what is variously known as 'platform capitalism' (Srnicek 2016), 'the platform society' (van Dijck et al. 2018), and 'the platformization of cultural production' (Nieborg and Poell 2018), but these pressures run counter to neoliberal economic policies that favour deregulation and austerity. The purpose of this chapter, which also serves as a conclusion to the book, is to view Unity's slogan—'democratizing game development'—as a policy discourse that emerges from, and is consonant with, the above developments. The discourse of democratization drives Unity and its core agenda, but it also performs political 'work' within videogame culture more broadly, where it is met with mixed feelings. The challenge for any piece of cultural software is to strategically manage, support, and govern these feelings in a way that serves to bolster the software's network effects.

Rather than asking whether Unity has truly democratized videogame development, we are more interested in the effects of Unity's policy discourse on the values, norms, and literacies upheld by videogame developers, critics, and players. Here, we are inspired by Nathaniel Tkacz's (2014) interrogation of the concept of 'openness' in relation to Wikipedia. As Tkacz (2014: 32) observes, openness is a concept 'whose meaning is so overwhelmingly positive it seems impossible even to question, let alone critique', to the extent that

> the open actively works against the development of a political language— if, that is, we take the political to extend beyond questions of just governance to the circulation and distribution of power and force, and take politics to mean the distributions of agency in general as well as the conflicts and issues that emerge when antagonistic flows intersect. (Tkacz 2014: 33)

Indeed, it is ironic that many of today's software companies—game engine providers among them—present themselves as politically neutral entities, given that they are very often the loci of our most fundamental political crises and power imbalances. As Tarleton Gillespie (2017a: n.p.) observes, the 'platform' metaphor, which is commonly used to describe the major software companies of our era (and strategically used by these companies to describe themselves), carries connotations of flatness, openness, and neutrality. This obscures the fact that platforms are deeply political entities that often mishandle their social and economic responsibilities. Unity is not the only software tool that claims to have democratized something—indeed, many cultural software utilize what we call a 'democratization *dispositif*' to justify their monopolizing and territorializing tendencies—but it is a useful case study for analysing how this *dispositif* can become mobilized as a means of generating power within a software ecology. For Unity, the discourse of democratization serves a particular socio-political purpose; it makes Unity seem normal, natural, or taken-for-granted. Put simply, it disarms critique. To this end, we follow Tkacz's (2014: 3) lead in arguing for 'a politics in the face of openness—a politics in spite of openness'.

HATRED OF DEMOCRATIZATION: FROM 'ASSET FLIPS' TO 'INDIEPOCALYPSE'

The industry and culture of videogames have a conflicted relationship with—and, in extreme cases, an underlying hatred of—democracy. In *Hatred of Democracy*, Jacques Rancière (2006) observes that efforts to uphold or justify democratic ideals and practices of equality are often premised on contradictory desires to undermine, govern, or weaponize democracy. In videogame culture, this contradiction plays out in unique ways. Videogame culture is, as Christopher A. Paul (2018) argues, largely built on a 'meritocratic' social order wherein raw talent and hard work are believed to hold sway when it comes to determining one's position in the social hierarchy. In a meritocracy, those who manage to ascend the social hierarchy are said to do so solely by virtue of their grit and determination, rather than by virtue of their social class, educational background, race, gender, or identity. Systemic power imbalances, institutional biases, and discriminatory practices are deemed irrelevant to one's ability to attain just reward for their hard work and effort. As Paul (2018) observes, meritocratic norms are very often replicated in the narratives and play structures of videogames themselves—many videogames are, for example, premised on some variation of the archetypical 'rags-to-riches' narrative, and place an extreme emphasis on technical skill, competency, and competition.

As flagged in previous chapters, meritocratic norms are also brought to bear upon the world of videogame production. Game engines such as Unity have recently come under fire for precisely the same reasons they are celebrated: for lowering the barrier of entry to becoming a videogame developer and for enabling a wider variety of development skills, practices, values, and subject positions to proliferate in videogame culture. Videogame culture's 'hatred' of democracy is rarely as explicit as the term hatred implies, but as Rancière (2006) maintains, hatred of democracy often expresses itself through a seemingly banal and contradictory desire to celebrate democracy while ensuring its supposed excesses are held in check. Hatred of democratization in videogame culture stems from a meritocratic belief that videogame development is a specialized craft that requires a certain degree of skill, knowledge, and hard work, rather than a field of cultural production that anyone can participate in.[1]

A recent example of this meritocratic ideology at work—and one that speaks directly to the mixed reception of Unity's policy discourse—is the growing backlash against videogames built from prefabricated assets. In 2017, videogame company Valve began cracking down on 'fake games' being sold on its online distribution platform, Steam. Valve's criteria for identifying fake games are somewhat ill-defined,[2] though key offenders include videogames made from prefabricated assets, downloaded from online asset stores such as Unity's Asset Store or Epic's Unreal Marketplace. Several YouTube pundits—perhaps the most prominent being Jim Sterling, whose videos were discussed in previous chapters—took to labelling these videogames asset flips. Asset flipping, as already defined, is a derogatory expression describing a videogame cobbled together from prefabricated assets and derivative design techniques, whose perceived main purpose is to turn a quick profit for its developers. Having sought consultancy with several high-profile YouTubers, Valve set out to revise its gatekeeping protocols and quality assurance

---

[1] This is not to say that Unity developers are unskilled; on the contrary, and as outlined in previous chapters, Unity development affords a wider range of creative skill sets to access the craft of videogame development than has traditionally been allowed.

[2] The company implemented a 'confidence metric' for gatekeeping purposes that, according to a May 2017 blog post, 'is built from a variety of pieces of data, all aimed at separating legitimate games and players from fake games and bots' (jonp 2017: n.p.).

baseline, with the aim of stemming the distribution of fake games and asset flips on its Steam webstore. Perhaps the most high-profile video-game to be accused of asset flipping is *PlayerUnknown's Battlegrounds* (*PUBG*; PUBG Corporation, 2017), an online 'battle royale' videogame where players are parachuted into multiplayer maps and pitted against each other in last-person-standing style deathmatches. *PUBG* was built using the Unreal engine, which, like Unity, has an asset store called the Unreal Marketplace. Some players noticed that several 3D models used in *PUBG*'s multiplayer maps bore a resemblance to certain prefabricated assets available on the Unreal Marketplace. Responding to criticism from online commentators, *PUBG*'s communications director, Ryan Rigley, defended his team's decision to use prefabricated assets. 'That's the only way you can spin out a game fast, and for a reasonable price', he wrote on the *PUBG* subreddit (PUBG_Riggles 2018: n.p.). Rigley quotes one of the videogame's lead artists as adding: 'Why should one of my artists spend two weeks on a generic sculpt if they could instead spend that two weeks adding real value for players elsewhere? How many times should a telephone booth be modelled? How many times do we gotta sculpt a cash register?' (PUBG_Riggles 2018: n.p.).

While the act of using store-bought assets instead of investing the resources and labour to produce bespoke ones is seen here by players as a sign of incompetency, laziness, or profiteering, Rigley's response instead positions it as a way to avoid unnecessary repetition so as to focus resources and labour towards aspects of development considered more crucial. John Vanderhoef (2019: n.p.) reconfigures the act of asset flipping as a more productive act of 'asset poaching', as a 'bricolage approach to gamemaking [that] privileges the accessibility of cultural production in the digital game space'. A high-profile Unity-developed videogame that relies heavily on poaching prefabricated assets—albeit one that prompted a very different kind of reaction in the videogame community—is *Getting Over It with Bennett Foddy* (Foddy, 2017). Launched on Steam in October 2017, *Getting Over It* has the player scaling a mountain of assorted objects and paraphernalia, using only a Yosemite hammer (see Fig. 6.1). *Getting Over It* is deliberately chal-lenging, even masochistic. Players are punished for making even minor mistakes, which occur frequently thanks to a somewhat unconventional physics system and sensitive mouse controls. Of note, however, is that *Getting Over It* deliberately embraces a 'fake game' (or, more accurately, 'b-game') aesthetic. The mountain players are tasked with 'getting over'

Fig. 6.1 Screenshot of *Getting Over It with Bennett Foddy* (Foddy, 2017). By permission of Bennett Foddy

is constructed from seemingly random art assets—household furniture items, gardening tools, construction site equipment, and so on—most of which were obtained for free from various asset stores; a handful of which were purchased; and an even smaller minority of which were custom-made. In the videogame's narrated commentary, Bennett Foddy, sole developer of *Getting Over It*, offers the following insight into his design philosophy:

> For years now, people have been predicting that games will soon be made out of prefabricated objects, bought from a store, and assembled into a world. And for the most part, that hasn't happened, because the objects in the stores are trash. I don't mean they look bad or that they're badly made, although a lot of them are. I mean they're trash in the way food becomes trash as soon as you put it in the sink [...] Over time, we've poured more and more refuse into this vast digital landfill we call the internet. It now vastly outnumbers and outweighs the things that are fresh and untainted and unused. When everything around us is cultural trash, trash becomes the new medium, the lingua franca of the digital age. You can build culture out of trash, but only trash culture: B-games, B-movies, B-music, B-philosophy. Maybe this is what digital culture is. A monstrous mountain of trash, the ash heap of creativity's fountain. A landfill of everything we ever thought of in it. Grand, infinite, and unsorted.

In an interview published on the website VentureBeat, Foddy reflects on the growing vitriol directed at poached assets. He argues that Valve's crackdown on 'fake games' risks further marginalizing 'people making games that just do not neatly fit into orthodox videogame genres', suggesting that 'a few of the people most vocally calling for quality control would not mind at all if those [unorthodox] games were eliminated from the market' (in Grubb 2018: n.p.).

The meritocratic norms informing attitudes towards asset poaching are also manifest in the 'indiepocalypse', discussed in previous chapters. The indiepocalypse anxiety of there being 'too many games and too many developers' is stoked by prophetic warnings of a videogame industry oversaturated with videogames made by small-to-medium-sized studios and indies struggling to make ends meet (see Wright 2018: n.p.). Several of our respondents linked this oversaturation to the apparent excesses of democratized videogame-making tools and distribution platforms. One of our respondents speculated that the 'accessibility or openness of engines' will be the 'biggest thing that disrupts the industry in the future', specifically in terms of 'the amount of people getting in to [videogame development] and how that's going to affect funding, how it's going to affect the amount of the games out there and markets being flooded or markets closing themselves off to be more curated'. These anxieties stem from the fact that, in recent years, videogame development has been restored[3] to something of a folk craft, akin to making music or writing poetry (cf. Yang 2017). Debates about there being 'too many games and too many developers' are symptomatic of a worldview that perceives videogames as economic objects that exist primarily for the purposes of satisfying consumer desires and generating revenue, as opposed to cultural objects that exist because people have the means—and the desire—to make them. It is also plausible to argue that the oversaturation of content can be attributed to the curatorial strategies (or lack thereof) of the 'cultural intermediaries' responsible for gatekeeping—Valve's Steam, Apple's App Store, Google's Play Store, and so on—rather than the tools that enable content to exist (see Parker et al. 2018: 1964). Nonetheless, it is not our intention to dismiss developers' concerns over the alleged 'indiepocalypse'. For many, the indiepocalypse represents a threat to livelihood—the struggle to strike success

---

[3] As discussed in Chapter 1, videogame development was, up until the 1990s, a hobbyist and oftentimes informal practice.

in a market that is undeniably competitive—rather than a reflection of meritocratic ideologies as such. That said, it is clear that many videogame developers, critics, and players are struggling to come to terms with the redistribution of power that may result from a more 'democratized' landscape of videogame development. Furthermore, the notion that this redistribution of power will make it more difficult to make videogames in an economically sustainable manner suggests that 'democratizing' videogame development also brings it in line with a creativity *dispositif* that normalizes self-governance, entrepreneurship, and unpaid creative labour (McRobbie 2016).

In part due to the growing popularity of game engines such as Unity, videogame development is no longer a niche sub-field of software development, but rather a field of cultural production that, as illustrated in the previous chapters, is articulated to multiple circuits of meaning-making. Like any field of cultural production, the videogame-making ecology consists largely of people who have little or no expectation of making money from the things they produce. Instead, developers are increasingly expected to find innovative, entrepreneurial, and personalized ways of developing their careers without relying on ongoing employment, welfare support, or even cultural policy aimed at promoting growth in the creative sector. It is in this environment that Unity's alleged commitment to democratization and equality takes on such an immense appeal, while simultaneously functioning as a source of anxiety for many.

UNITY'S DEMOCRATIZATION *DISPOSITIF*

People feel empowered by Unity not only because of the tools it provides, but also because it appears to make a sustained, policy-driven commitment to democracy and equality in a political environment where such a commitment is typically felt to be lacking. There are multiple theories explaining Unity's rapid ascension to a quasi-monopolistic dominance, but one, perhaps under-acknowledged explanation is that the company has leveraged the symbolic power of democratization at a time when developers face increased vulnerability and precariousness because of, for example, the erosion of state welfare. This can be seen not only in the engine's accessibility, or even in the company's decision to provide 'free' licences for its core software, but also in its efforts to cultivate an affective community—one that builds on long-standing 'communitarian practices' in the independent development communities

(Guevara-Villalobos 2011: 3)—through extensive investment in education, events, and support. Ultimately, in Morgan's words, Unity carries
the promise that a videogame 'built by two, four, five people [can com-
pete] on Steam alongside games that were built by 400 or 500 people'.
Unity's commitment to accessibility, low-cost software, and affective
community creates an affective space where developers are granted a slight
degree of social security to explore possibilities for self-entrepreneurship
in what would otherwise be a career path fraught with risk and uncertainty. This creates what Angela McRobbie (2016) calls a 'creativity *dispositif*. The creativity *dispositif* refers to the 'toolkits, instruments and new
entrepreneurial pedagogies' that encourage prospective creative workers
to pursue work in the media and design sectors (McRobbie 2016: 86). In
McRobbie's (2016: 34–35) words, the creativity *dispositif*

> oversees novel forms of job creation (in times of both unemployment
> and under-employment), the defining features of which are imperma-
> nent, short-term, project-based or temporary positions; it orchestrates an
> expansion of the middle classes in the light of the policies adopted by most
> national governments in recent years to increase the numbers of students
> attending universities and art colleges and at the same time it supports
> the creative activities of this arriviste middle class, allowing them to act as
> guinea pigs for testing out the new world of work without the full raft of
> social security entitlements and welfare provision that have been associated
> with the post-Second World War period.

The creativity *dispositif* supports a new kind of ethos for creative
workers: one where people are encouraged to adopt strategies of
self-entrepreneurship in the absence of welfare support; where, in one
respondent's words, user-friendly software tools 'give everyone the opportunity to create value where there was no value before'; and where higher
education is restructured to help students find their niche in the risk economy of creative work. Riffing on McRobbie's term, Unity Technologies's
governance structure can be understood as a democratization *dispositif*
for the prospective Unity developer. The very notion of 'democracy' is, as
political philosopher Ernesto Laclau (2007 [1996]) establishes, an 'empty
signifier' capable of facilitating a multiplicity of contradictory demands
and subject positions, meaning that it can be deployed in almost any discursive context to galvanize support and achieve a hegemony. As discussed
earlier, software companies often deploy terms such as 'democratization'

and 'openness' in their marketing strategies for these exact purposes (Tkacz 2014). As several of our respondents made clear, democracy can be a sensitive topic in a cultural context where an underlying 'hatred of democracy' (Rancière 2006) has taken root in ostensibly democratic societies. For Unity, 'democratization' facilitates a multiplicity of promises—the promise of creative freedom, self-publishing, community, openness, and entrepreneurship—that together function to make videogame development seem like an accessible and exciting career path.

Democratization is a powerful, mobilizing, and oftentimes-sensitive concept in videogame development (see Harvey 2014). Whenever democratization came up in our interviews—and it often did, unprompted—its status as an empty signifier became clear. Respondents linked democratization to accessibility; versatility; empowerment; a diversification of content and culture; a recognition of the labour of artists and designers as opposed to just programmers; self-publishing; interfaces that are 'creative-friendly' and 'visual'; the open sharing of knowledge; freedom to create and sell products on the Asset Store; and care and community. Yet, democratization was also linked to contradictory feelings of 'a lot more content and a lot more competition'; homogenized design practices; the threat of monopolization; Silicon Valley 'blue-sky' ideology; and the ongoing presence of 'computer-science baggage' in ostensibly creative-friendly design tools. One participant even went so far as to describe Unity's bid to democratize development as 'absurd':

> Democracy would imply that the wider world would decide who's running Unity, but they don't [...] I don't think there's a purer example of [hierarchy] than in software development [...] If your project is big enough, you have a tree structure of people in charge [...] So, with Unity, are [Unity's key stakeholders] saying that if people start complaining about them, that they can just be voted out of the company? Of course they're gonna maintain control of Unity. They invented it, they run it. I don't really understand what democratizing game development is meant to mean.

Respondents often prefaced these and other concerns by first signalling their faith in Unity's policy discourse. Most of our respondents—from students to industry professionals—regarded Unity Technologies's commitment to democratization not simply as a marketing slogan, but rather as an underlying political orientation that aligned with their personal beliefs, values, and subject positions. Several respondents expressed concern that being critical of Unity would come off as overly negative, given the company's commitment to democratization and its covenant of good faith with the community.[4] In this way, Unity's democratization *dispositif* can be understood as a kind of governance mechanism; one that expands the engine's ecology to encompass not only a suite of software tools but also a collection of positive thoughts, feelings, and affects that, once enclosed in the system of governance, can be converted into capital.

As discussed throughout this book, Unity's bid to democratize development is consonant with a broader neoliberalization of work and subjectivity in digital culture. For Chun (2011: 8), the discourses of empowerment that underpin software culture—for example, Unity's discourse of democratization or Wikipedia's discourse of openness—are premised on the notion that 'the worker does not simply own his/her labor, but also possesses his/her own body as a form of "human capital"'. Cultural software enrol media creatives in social/affective eco-systems where people are encouraged to become entrepreneurs of themselves (cf. Foucault 2008: 226) and where creative workers become 'highly reliant on informal networking but without the support of those underpinned by any institutional "trade association"' (McRobbie 2002: 519). Along these lines, one of our respondents described game engines as 'platforms for anyone to make a living, sharing pieces of work'. As he put it,

> You could look at this game engine thing as creating new parts of the economy. Because, you know, the economy is human labour. Money is created from the value people bring. And to make value, you need an opportunity. And these game engines give everyone the opportunity to create value where there was no value before.

According to a 2017 'U.S. Emerging Jobs Report' published by LinkedIn, 'Unity developer' was the 7th most in-demand job title among North American employers, and the only entry among the top-ten searches identified to feature a company's name (Economic Graph Team 2017). Referring to this statistic in a keynote address given at

---

[4]This suppression of 'bad affect' is a hallmark of creative entrepreneurship in neoliberal capitalism, where, as McRobbie (2016: 25) puts it, 'presentation of self is incompatible with a contestatory demeanor. Personal angst, nihilism or mere misgivings [...] must be privately managed and, for the purposes of club sociality, carefully concealed'.

the 2018 Game Developers' Conference (GDC) in San Francisco to a room full of Unity developers, John Riccitiello remarked that 'we live in a time when there are all sorts of theories about what creates employment. And my answer is: it's you' (Unity 2018: n.p.). Democracy is a term that has been 'dearticulated' from political discourses and 'rearticulated' in software culture to neoliberal discourses of self-entrepreneurism, self-sovereignty, and self-governance (see Hall 1986; cf. D'Acci 2004: 435–435). In the process, the hegemonic meaning of democracy has shifted, such that 'democratization' has become a signifier that naturalizes neoliberal modes of work and identity formation. When Riccitiello celebrates an audience of Unity developers as, first and foremost, 'job creators', he personifies Unity's democratization *dispositif* that rationalizes and individualizes cultural work under neoliberal capitalism and which frames corporate strategies of enrolment as social benevolency.

Democratization Beyond Unity

There is no denying that Unity has made the craft of videogame development more accessible, and that it is contributing to a diversification of the culture and industry of videogames. Unity has provided a gateway to 3D videogame development in a way that no other software tool has done previously. Many amateur and independent developers have utilized the 'free' resources provided by Unity to rethink how videogames can be made and played. Nonetheless, while Unity has provided a platform for diversity and accessibility in videogame development, it is important to recognize that diversity and accessibility do not exist *because of* Unity. Cultural theorists have routinely dismissed this kind of thinking as technologically deterministic. Fringe development communities and cultures have always found ways to make videogame development more accessible, diverse, and equitable. In fact, if Unity can be said to have democratizing effects, then these effects are only possible through what Paul du Gay et al. (1997: 52) call an 'articulation of production to consumption'—that is, Unity's toolset only *becomes* democratizing once it is brought into contact with the productive energies of its collective user base. Indeed, as discussed in Chapter 2, Unity's very conditions of existence are predicated on long-standing (albeit largely informal) histories of hobbyist, modding and grassroots videogame-making tools and practices.

Moreover, while Unity may have some claim of legitimacy in making the tools of videogame development more accessible, it is important to clarify the limits of its supposedly democratizing effects. Unity has not democratized employment opportunities, nor has it addressed tendencies within studio environments to turn a blind eye towards issues of precariousness, toxicity, burnout, and exploitation; the same goes for the industry's long-standing ambivalences towards collective organisation. Granted, Unity's mission statement is not 'democratizing workplace politics', but it is worth drawing attention to the neoliberal values underpinning its specific vision of democratization. Unity's democratization *dispositif* displaces the task of labour reform onto self-governing Unity users, rather than the dominant market actors and institutional structures within whose remit these power imbalances first originated. Moreover, Unity Technologies's long-term business plan is opaque, meaning that it is difficult to get a sense of when and how its business model will change and what the ramifications of these changes will be for its users. At present, Unity's software and licensing structures are subject to frequent updates—updates that can have significant ramifications for its user base. As established, this user base consists not only of videogame creators but also of Asset Store developers and contributors to community forums, whose support labour is the lifeblood of the engine. We do not want to suggest that these developers have been somehow 'duped' into using the tool, or that they do not have a critical grasp on their own videogame-making practices. On the contrary, our respondents were often well aware of the critical issues raised throughout this book, but nonetheless viewed Unity as a significant step forward from the proprietary engines that once stifled the field of production.

Before Unity established its democratization *dispositif*, there already existed a developer counterculture that, as discussed in Chapter 2, had made long-standing aggravations for 'democratization' in the culture and industry of videogames, if we understand democratization to refer broadly to practices of equality. Members of this counterculture sought to overcome the technological barriers to videogame development and, in the process, to question the institutional logics of the triple-a industry. Since that time, however, the concept of democratization has been dearticulated from its countercultural origins and absorbed into the democratization *dispositif*, where it now functions as a policy discourse for companies such as Unity Technologies. However, it is important to acknowledge that videogame-making countercultures continue to produce new kinds of technologies, techniques, and subjectivities that elude, confound, or occasionally short-circuit (that is, draw productive energy from) dominant platform and software ecologies. These include, for example, various grassroots game engines and videogame-making communities, such as those associated with Twine, Bitsy, and Pico-8. As discussed in previous chapters, these grassroots engines are created and maintained by individuals and communities, and either explicitly or implicitly resist the typical logics of governance and control that engines such as Unity hold over videogame development cultures. Their creators are rarely profiting from the engines and are often reliant on the engine's community for ongoing support, resources, and maintenance.

Bitsy, for instance, is a small game engine by Adam Le Doux that facilitates the creation of small, 2D, story-driven videogames. Its active online community develops expansions for the engine (including features such as audio implementation), produces zines of new videogame releases, maintains a wiki, and holds monthly 'game jams' to prompt the creation of more Bitsy games. If development and distribution platforms such as Unity, Unreal, and Steam work to homogenize, govern, and regulate particular aspects of the formalized videogame industry, then grassroot engines imagine a plurality of communal production cultures at the fringes of the videogame industry beyond these jurisdictions. Like Unity, grassroots game engines afford specific workflows and grains, as discussed in Chapter 3. However, whereas the dominant engines strive for a perception of neutrality where they are 'rewarded for facilitating expression but not liable for its excesses' (Gillespie 2010: 356), grassroots engines are explicitly non-neutral in their politics and affordances. They do deliberately what dominant engines pretend not to: promote particular ways of making videogames and being videogame-makers. Grassroots engines provide alternative spaces of affective intermediation, offering open, community-driven toolsets that utilize the 'platform logic' for expressive, communal, and subversive (rather than monopolistic) purposes. It is also important to note that these grassroots engines are, on the whole, accessible to non-programmers and everyday creators in a way that even Unity is not. As discussed in Chapters 4 and 5, Unity is experienced as democratizing primarily for those who already possess an understanding of videogames, videogame culture, and the language of computing—what Graeme Kirkpatrick (2013: 72) calls the 'gamer habitus'—rather than more interdisciplinary creatives who might opt to use Unity as one among many cultural software tools.

The growth of grassroots engines is proof that the people who comprise videogame-making communities are more than capable of shaping and, indeed, determining social, cultural, and technological change, without the bestowing of democracy from a company such as Unity Technologies. In a short manifesto titled 'KILL UNITY; WE ARE ENGINES', videogame developer and academic Robert Yang (2018: n.p.) writes,

> No one way of making and doing cultural work should have such a monopoly and stranglehold on an entire creative community [...] This [manifesto] is a call to build more tools, more frameworks, more engines, more ways of doing and thinking about games and play [...] Why can't you paint a game by singing, why can't you breed and mutate 3D models, why can't you sculpt an AI? [...] Kill Unity, death to Unreal; for we are engines.

Although Yang's (2018: n.p.) manifesto is deliberately tongue-in-cheek—he notes that he '[doesn't] actually want to kill anyone, and [has] been using Unity for years'—it touches on a fundamental truism: *in software culture, bodies—rather than tools—are engines of capital and creativity.* The alleged 'democratization' of videogame development is premised on a powerful belief in technological determinism—that is, the notion that tools, rather than people, are agents of social change. Yet, Unity's success as a low barrier to entry videogame-making tool is less a result of its technical affordances and more of its ability to strategically build on (and, in the process, to interpellate and erase [Vogel 2017]) existing grassroots tools, practices, and communities. Moreover, as discussed in Chapter 2, one of the main reasons Unity has been able to rapidly scale up is because of the (oftentimes invisible) support labour of its various developer communities, whose members have contributed assets and plugins to the Asset Store, formed online communities, and created online tutorials for inexperienced users. The circuits of cultural software associated with Unity function to configure and mobilize these spaces of affective intermediation and make them serviceable and legible to Unity's platform ecology. Yet, these tools, practices, and communities existed long before Unity and will continue to exist after Unity. The challenge for Unity developers, then, is to realize the democratizing potentials latent in their own videogame-making activities, which are articulated, but not reducible, to the Unity engine.

CONCLUSION

Just as game engines in their myriad of commercial, proprietary, and grassroots forms have come to underpin videogame production, cultural software more broadly have come to be, as Manovich (2013) argues, 'engines' of cultural production in the twenty-first century. Just as cultural theorists of the twentieth century rendered legible the mediating impact of material creative tools such as musical instruments, film stocks, and paint types on the field of cultural production, it is crucial for us to develop ways to comprehend the cultural software that underpin contemporary cultural production, without reducing them to either deterministic actors, romantic enablers, or passive platforms. In this book, we have focused specifically on the Unity game engine, tracing its cultural software circuits so as to articulate a range of tensions and movements within the videogame field in particular. However, one could just as easily look at the circuits of a different piece of cultural software. For instance, one could consider Photoshop's dominance in the graphic design field, its impact on (and emergence from) the field of photography, and how conflicting ideals of Photoshop are mobilized in discourses around, for instance, fake news or representations of women. Alternatively, one might use this framework to consider the mediating impact of Microsoft's Powerpoint slideshow software on pedagogical practices. In essence, cultural software are more than simply programs running on computers. They are cultural entities that configure complex circuits—circuits that inscribe particular ways of working, creating, and knowing in any given field of cultural production. Ultimately, they enrol constituents into governed ecosystems—ecosystems that obscure their governing through discourses of democratization, empowerment, openness, and user-friendliness.

REFERENCES

Chun, Wendy Hui Kyong. 2011. *Programmed Visions: Software and Memory.* Cambridge: MIT Press.

D'Acci, Julie. 2004. "Cultural Studies, Television Studies, and the Crisis in the Humanities." In *Television After TV: Essays on a Medium in Transition,* edited by Lynn Spigel and Jan Olsson, 418–445. Durham: Duke University Press.

du Gay, Paul, Stuart Hall, Linda Janes, Hugh Mackay, and Keith Negus. 1997. *Doing Cultural Studies: The Story of the Sony Walkman.* London: Thousand Oaks.

Economic Graph Team. 2017. "LinkedIn's 2017 U.S. Emerging Jobs Report." Linkedin, December 7. https://economicgraph.linkedin.com/research/LinkedIns-2017-US-Emerging-Jobs-Report.

Foucault, Michel. 2008. *The Birth of Biopolitics: Lectures at the Collège de France, 1978–79*. Basingstoke: Palgrave Macmillan.

Gillespie, Tarleton. 2010. "The Politics of 'Platforms'." *New Media & Society* 12 (3): 347–364.

Gillespie, Tarleton. 2017a. "The Platform Metaphor, Revisited." Culture Digitally, August 24. http://culturedigitally.org/2017/08/platform-metaphor/.

Gillespie, Tarleton. 2017b. "Governance of and by Platforms." In *SAGE Handbook of Social Media*, edited by Jean Burgess, Thomas Poell, and Alice Marwick, 254–278. Los Angeles: SAGE Publications.

Grubb, Jeff. 2018. "In Defense of Asset Flips on Steam." VentureBeat, July 12. https://venturebeat.com/2018/07/12/in-defense-of-asset-flips-on-steam/.

Guevara-Villalobos, Orlando. 2011. "Cultures of Independent Game Production: Examining the Relationship Between Community and Labour." In *Proceedings of DiGRA 2011 Conference: Think Design Play*, 1–18.

Hall, Stuart. 1986. "On Postmodernism and Articulation: An Interview with Stuart Hall." *Journal of Communication Inquiry* 10 (2): 45–60.

Harvey, Alison. 2014. "'Twine' Revolution: Democratization, Depoliticization, and the Queering of Game Design." *Game* 3: 95–107.

jonp. 2017. "Changes to Trading Cards." Blog post, May 17. https://steamcommunity.com/games/593110/announcements/detail/1954971077935371954.

Kirkpatrick, Graeme. 2013. *Computer Games and the Social Imaginary*. Cambridge: Polity Press.

Laclau, Ernesto. 2007 (1996). *Emancipation(s)*. London: Verso.

Manovich, Lev. 2013. *Software Takes Command*. Cambridge: MIT Press.

McRobbie, Angela. 2002. "Clubs to Companies: Notes on the Decline of Political Culture in Speeded up Creative Worlds." *Cultural Studies* 16 (4): 516–531.

McRobbie, Angela. 2016. *Be Creative: Making a Living in the New Culture Industries*. Cambridge: Polity Press.

Nieborg, David B., and Thomas Poell. 2018. "The Platformization of Cultural Production: Theorizing the Contingent Cultural Commodity." *New Media & Society* 20 (11): 4275–4292.

Parker, Felan, Jennifer R. Whitson, and Bart Simon. 2018. "Megabooth: The Cultural Intermediation of Indie Games." *Games and Culture* 20 (5): 1953–1972.

Paul, Christopher A. 2018. *The Toxic Meritocracy of Video Games: Why Gaming Culture Is the Worst*. Minneapolis: University of Minnesota Press.

Plantin, Jean-Christophe, Carl Lagoze, Paul N. Edwards, and Christian Sandvig. 2018. "Infrastructure Studies Meet Platform Studies in the Age of Google and Facebook." *New Media & Society* 20 (1): 293–310.

PUBG_Riggles. 2018. "Re: Brendan Greene Wants to Kill Me." Online forum comment. https://www.reddit.com/r/PUBATTLEGROUNDS/comments/8qu07m/brendan_greene_wants_to_kill_me/e0ma17n/. Accessed 19 October 2018.
Rancière, Jacques. 2006. *Hatred of Democracy.* London: Verso.
Srnicek, Nick. 2016. *Platform Capitalism.* Malden: Polity Press.
Tkacz, Nathaniel. 2014. *Wikipedia and the Politics of Openness.* Chicago: The University of Chicago Press.
Unity. 2018. "Unity at GDC Keynote—March 19, 2018." YouTube, March 19. https://www.youtube.com/watch?v=cmRSkHl-Gv0.
Vanderhoef, John. 2019. "Throwing Shit at the Wall: Maligned Aesthetics, Asset Flipping, and the Politics of Value in Informal Game Development." In *Presented at Society of Cinema and Media Studies Annual Conference, Seattle,* March 13–17.
van Dijck, José, Martijn de Waal, and Thomas Poell. 2018. *The Platform Society: Public Values in a Connective World.* New York: Oxford University Press.
Vogel, Michael. 2017. "Japanese Independent Game Development." MA dissertation, Georgia Institute of Technology. https://smartech.gatech.edu/bitstream/handle/1853/58640/VOGEL-THESIS-2017.pdf?sequence=1&isAllowed=y.
Wright, Steven. 2018. "There Are Too Many Video Games. What Now?" Polygon, September 28. https://www.polygon.com/2018/9/28/17911372/there-are-too-many-video-games-what-now-indiepocalypse.
Yang, Robert. 2017. "Lol We're All Poor." *Radiator,* June 26. https://www.blog.radiator.debacle.us/2017/06/lol-were-all-poor.html.
Yang, Robert. 2018. "KILL UNITY; WE ARE ENGINES." Itch.io, February 12. https://radiatoryang.itch.io/kill-unity.

INDEX

© The Editor(s) (if applicable) and The Author(s), under exclusive license   119
to Springer Nature Switzerland AG 2019
B. Nicoll and B. Keogh, *The Unity Game Engine and the Circuits of Cultural Software*, https://doi.org/10.1007/978-3-030-25012-6